Official

# Higher Biology

2010–2014



HODDER GIBSON
AN HACHETTE UK COMPANY

Hodder Gibson is grateful to the copyright holders, as credited on the final page of the Question Section, for permission to use their material. Every effort has been made to trace the copyright holders and to obtain their permission for the use of copyright material. Hodder Gibson will be happy to receive information allowing us to rectify any error or omission in future editions.

Hachette UK's policy is to use papers that are natural, renewable and recyclable products and made from wood grown in sustainable forests. The logging and manufacturing processes are expected to conform to the environmental regulations of the country of origin.

Orders: please contact Bookpoint Ltd, 130 Park Drive, Abingdon, Oxon OX14 4SE. Telephone: (44) 01235 827720. Fax: (44) 01235 400454.

Lines are open 9.00–5.00, Monday to Saturday, with a 24-hour message answering service. Visit our website at www.hoddereducation.co.uk. Hodder Gibson can be contacted direct on: Tel: 0141 848 1609; Fax: 0141 889 6315; email: hoddergibson@hodder.co.uk

This collection first published in 2014 by

Hodder Gibson, an imprint of Hodder Education,

An Hachette UK Company

2a Christie Street

Paisley PA1 1NB

BrightRED PUBLISHING Hodder Gibson is grateful to Bright Red Publishing Ltd for collaborative work in preparation of this book and all SQA Past Paper, National 5 and Higher for CfE Model Paper titles 2014.



Typeset by PDQ Digital Media Solutions Ltd, Bungay, Suffolk NR35 1BY

Printed in the UK

A catalogue record for this title is available from the British Library

ISBN 978-1-4718-3676-3

3 2 1

2015 2014

# Introduction

## Study Skills – what you need to know to pass exams!

### Pause for thought

Many students might skip quickly through a page like this. After all, we all know how to revise. Do you really though?

*Think about this:*

"IF YOU ALWAYS DO WHAT YOU ALWAYS DO, YOU WILL ALWAYS GET WHAT YOU HAVE ALWAYS GOT."

Do you like the grades you get? Do you want to do better? If you get full marks in your assessment, then that's great! Change nothing! This section is just to help you get that little bit better than you already are.

There are two main parts to the advice on offer here. The first part highlights fairly obvious things but which are also very important. The second part makes suggestions about revision that you might not have thought about but which WILL help you.

### Part 1

DOH! It's so obvious but ...

*Start revising in good time*

Don't leave it until the last minute – this will make you panic.

Make a revision timetable that sets out work time AND play time.

*Sleep and eat!*

Obvious really, and very helpful. Avoid arguments or stressful things too – even games that wind you up. You need to be fit, awake and focused!

*Know your place!*

Make sure you know exactly **WHEN and WHERE** your exams are.

*Know your enemy!*

**Make sure you know what to expect in the exam.**

How is the paper structured?

How much time is there for each question?

What types of question are involved?

Which topics seem to come up time and time again?

Which topics are your strongest and which are your weakest?

Are all topics compulsory or are there choices?

*Learn by DOING!*

There is no substitute for past papers and practice papers – they are simply essential! Tackling this collection of papers and answers is exactly the right thing to be doing as your exams approach.

### Part 2

People learn in different ways. Some like low light, some bright. Some like early morning, some like evening / night. Some prefer warm, some prefer cold. But everyone uses their BRAIN and the brain works when it is active. Passive learning – sitting gazing at notes – is the most INEFFICIENT way to learn anything. Below you will find tips and ideas for making your revision more effective and maybe even more enjoyable. What follows gets your brain active, and active learning works!

*Activity 1 – Stop and review*

Step 1

When you have done no more than 5 minutes of revision reading STOP!

Step 2

Write a heading in your own words which sums up the topic you have been revising.

Step 3

Write a summary of what you have revised in no more than two sentences. Don't fool yourself by saying, "I know it, but I cannot put it into words". That just means you don't know it well enough. If you cannot write your summary, revise that section again, knowing that you must write a summary at the end of it. Many of you will have notebooks full of blue/black ink writing. Many of the pages will not be especially attractive or memorable so try to liven them up a bit with colour as you are reviewing and rewriting. **This is a great memory aid, and memory is the most important thing.**

7 Some candidates use a highlighter to remind themselves of important points they read within a question – this is very good practice.
8 We see lots of candidates getting marks for numerical answers – they clearly use their calculators effectively. Remember, most numerical answers are limited to numbers with a maximum of two decimal places.
9 We see very few blanks left in question papers, which is a great technique. Have a go using the words in the question – you never know!
10 Finally – candidates' handwriting is usually good. It is in their own interest that their handwriting is clearly legible – make sure yours is too!

## Where candidates sometimes get it wrong

There are techniques and practices which some candidates clearly miss out on and certain mistakes are very commonly seen. The points below are hints on how to maximise the marks you score.

1 Make sure you finish the paper – spend a maximum of 30 minutes on Section A, about 90 minutes on Section B and leave about 30 minutes for Section C.
2 Use the language of the question in your answers if you can – it will help keep you on the right track.
3 Marks can be lost for failure to use the accepted biological language which is found in the Arrangements document – it is safer to use terms from there in your answers. You should have a copy of the Arrangements – it can be downloaded from the SQA website.
4 Check the mark allocation and support lines in each question you attempt – if there is more than one mark there will be more than one support line and you will need to make more than one point.
5 If a question asks you to **state**, **name** or **give** in a question, then it's likely that only a short answer is needed. On the other hand, if you are asked to **describe** or **explain** something, you will need a longer answer.
6 In some questions, the wording is crucial and often contains a piece of information which is vital in reaching the correct answer. This is very often the case in the data question – always read the question a couple of times.
7 There will be a few words printed in **bold**.Take careful note of these, because the bold word will be very important in finding your answer.
8 You will always be asked to draw a graph – make sure you choose scales to fill most of the graph paper and use a ruler to connect the points in a line graph.
9 When doing calculations use the space for calculation provided – you will probably need it, and markers look at working in case it deserves a mark, even if the final answer is wrong.
10 Avoid using words like **it** and **they** – make sure you say what **it** actually is or what **they** actually are.
11 There is always a practical question. Make sure you know what a variable is, why controls are needed and how reliability can be improved in an investigation. Take a few minutes to read the question and try to visualise what is actually happening – a scribbled drawing on your scrap paper can help.
12 Candidates sometimes try to draw conclusions when they are simply asked for a description of a graph. A description only uses values to describe the trends in the data but does not attempt to explain them in any way.
13 When describing a graph line, ensure that you quote data points and units. In comparing graphical investigation results, it is good practice to use comparative terms such as **lower** or **higher** rather than absolute terms such as **low** or **high**.
14 When describing changes in tabulated data, always quote values to make it clear where trends in the data change.
15 Many good candidates lose marks for the use of words such as **never** and **always** – it's better to consider if words such as **sometimes**, **usually** or **often** might be better. Don't be tempted to say that an adaptation can **stop** or **prevent** competition – it's better to say it **reduces** competition. Don't be tempted to say that corrective mechanisms **stop** skin sweating – it's better to say sweating is **reduced**.
16 There are several metabolic pathways which are important in Higher Biology. Make sure you are able to give the consequences of a blocked step in a metabolic pathway shown in a question.
17 Many candidates have trouble with chromosome numbers in cells. Most cells have the diploid chromosome number, meaning they have two sets of chromosomes, but gametes are usually haploid, with only one set. Be careful about quoting the human chromosome complement – not all organisms are human!
18 It is easy to lose marks when notating the symbols of sex linked alleles – remember the allele must be associated with the sex chromosome on which it's found using a super-script, for example, the genotype of a male with the sex-linked allele R would be notated as $X^RY$.
19 If you run out of space for your answer, or want to make a change, ensure that you note down where you have put your changed answer, especially if it has had to go at the back of the booklet. Remember, there will always be a second graph grid at the back of the booklet if you need it. It's difficult to make changes to an incorrectly drawn graph – better score through and start again.
20 Section C is usually very challenging for candidates – 10 marks worth of extended writing is tough for anyone. Here are a few tips:
   - The two questions in Section C offer a choice – ensure that you give yourself a few minutes to make your choice. This is crucial because nearly every year one of the choices turns out to be a bit easier that the other. Note down the key words for each option – which seems more likely to yield marks?
   - Always write the title above your answer – it can sometimes contribute to your answer.
   - Diagrams are great for extended responses because they can be quicker than writing things out if you are short of time. Remember that diagrams must be labelled and, where there are arrows, the arrowheads must be clear and pointing in the right direction.
   - In Question 2, there is a mark for coherence and another for relevance – don't waste these! Coherence is about writing logically or in sections. The question title might suggest that there are two sections to the answer – make sure you separate your answer into the two parts and give each a short title. Relevance is really about not being irrelevant – if the question is about aerobic respiration, don't write anything about anaerobic!

## Good luck!

Remember that the rewards for passing Higher Biology are well worth it! Your pass will help you get the future you want for yourself. In the exam, be confident in your own ability. If you're not sure how to answer a question, trust your instincts and just give it a go anyway. Keep calm and don't panic! GOOD LUCK!

# HIGHER

# 2010

**12.** In mice, coat length is determined by the dominant allele **L** for long coat and the recessive allele **l** for short coat.

Coat colour is determined by the dominant allele for brown colour **B** and recessive allele for white colour **b**.

The genes are not linked.

What proportion of the offspring produced from a cross between two mice heterozygous for coat length and colour would have short brown coats?

A 1 in 16

B 3 in 16

C 9 in 16

D 1 in 4

**13.** Cystic fibrosis is an inherited condition caused by a recessive allele. The diagram below shows a family tree with affected individuals.

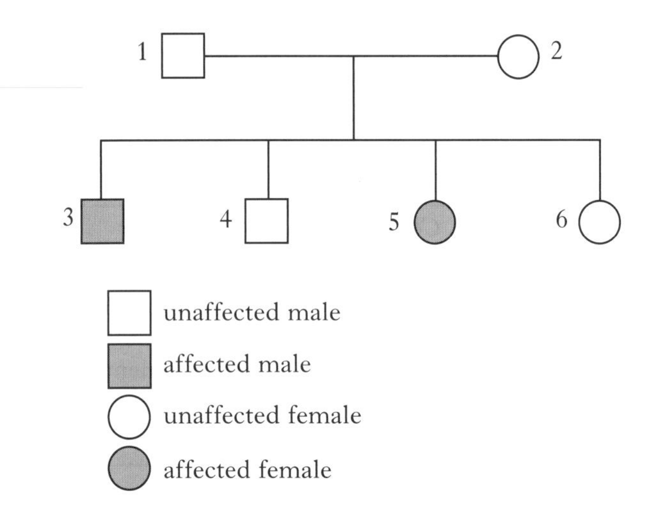


Which individuals in this family tree **must** be heterozygous for this condition?

A 3 and 5

B 4 and 6

C 1 and 2

D 2 and 6

**14.** In *Drosophila*, wings can be straight or curly and body colour can be black or grey.

Heterozygous flies with straight wings and black bodies were crossed with curly-winged and grey bodied flies.

The following results were obtained.

| *Number* | 797 | 806 | 85 | 89 |
|---|---|---|---|---|
| *Phenotype* | straight wings and black bodies | curly wings and grey bodies | straight wings and grey bodies | curly wings and black bodies |

These proportions of offspring suggest that

A genes for body colour and wing shape are on separate chromosomes

B crossing over has caused linked genes to separate

C the genes show independent assortment

D the genes must be carried on the sex chromosomes.

**15.** The diagram below represents the evolution of bread wheat. The diploid chromosome numbers of the species involved are given.

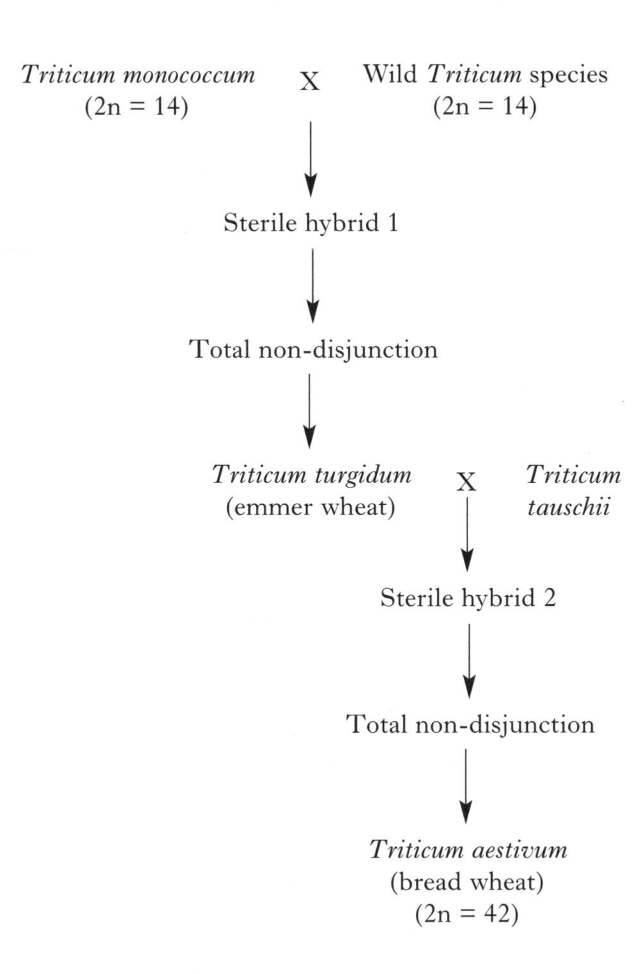


Which line in the table below identifies correctly the diploid chromosome numbers of *Triticum turgidum* (emmer wheat) and *Triticum tauschii*?

| | *Triticum turgidum* | *Triticum tauschii* |
|---|---|---|
| A | 14 | 14 |
| B | 28 | 14 |
| C | 14 | 28 |
| D | 28 | 28 |

**16.** The flow chart below represents the programming of *E. coli* bacteria to produce human insulin.

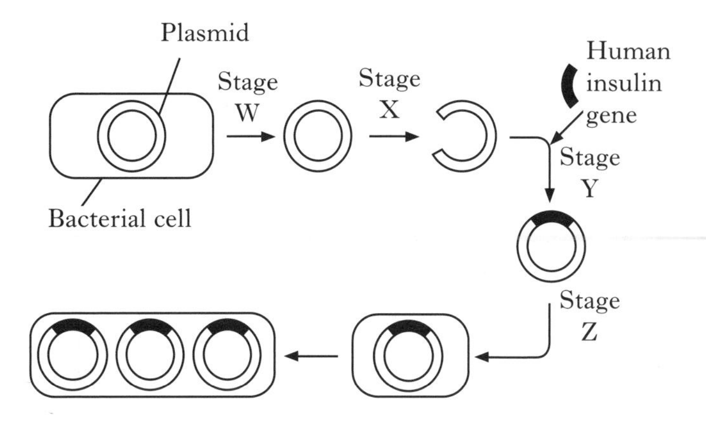


Which line in the table below identifies correctly the stages at which an endonuclease and a ligase are used?

| | *Endonuclease* | *Ligase* |
|---|---|---|
| A | Stage X | Stage W |
| B | Stage Y | Stage Z |
| C | Stage X | Stage Y |
| D | Stage Y | Stage X |

**17.** Which of the following statements could be **true** of cooperative hunting?

1 Individuals gain more energy than from hunting alone.

2 Both dominant and subordinate animals benefit.

3 Much larger prey may be killed than by hunting alone.

A 1 and 2 only

B 1 and 3 only

C 2 and 3 only

D 1, 2 and 3

**[Turn over**

**18.** The table below shows the mass of water gained and lost by a small mammal over a 24-hour period.

| | *Mass of water lost or gained* (g) |
|---|---|
| Food | 6 |
| Metabolic water | 54 |
| Exhaled air | 45 |
| Urine | 12 |
| Faeces | 3 |

What percentage of water gained comes from metabolic water?

A 9%

B 45%

C 54%

D 90%

**19.** The graph below shows the relationship between the ratio of body masses of two male fish and the average time they spend fighting for territory.

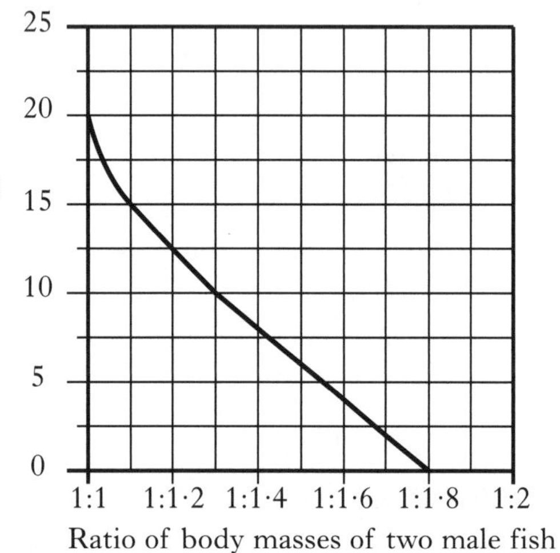


For how long will a fight between two fish, weighing 6 g and 9 g respectively, be expected to last?

A 6 minutes

B 10 minutes

C 15 minutes

D 17 minutes

**20.** Increased grazing by herbivores in a grassland habitat can result in an increase in the number of different plant species present in the habitat.

This is because

A some plants tolerate grazing because they have low meristems

B damage to dominant grasses by grazing allows the survival of other species

C grasses can regenerate quickly following damage by herbivores

D many grassland species produce toxins in response to grazing.

**21.** The diagram below shows a section of a woody twig.

Which is a region of summer wood?

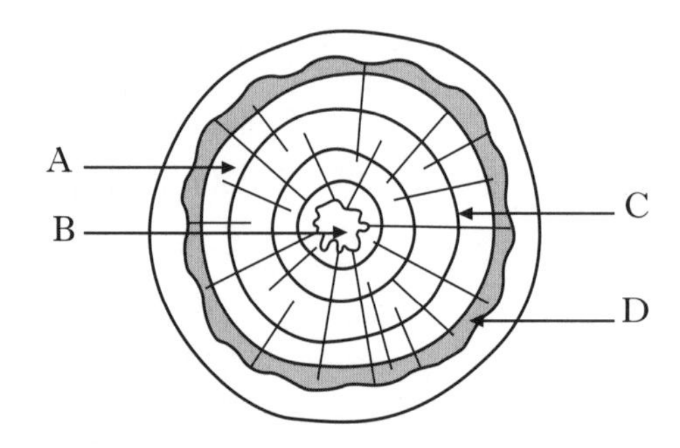


**22.** According to the Jacob-Monod hypothesis, a regulator gene is responsible for

A coding for the production of an inducer molecule

B switching on an operator

C coding for the production of a repressor molecule

D switching on a structural gene.

**23.** The graph below shows the effect of photoperiod on the onset of flowering in a species of plant.

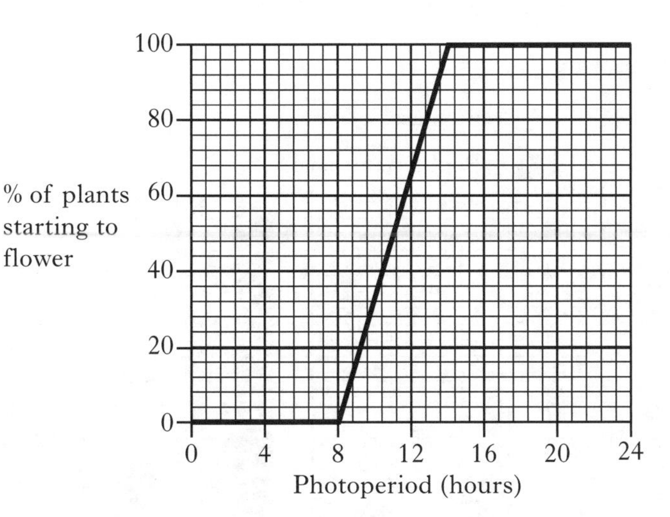


The graph shows that the plant is a

A long day species with a critical photoperiod of 8 hours

B long day species with a critical photoperiod of 14 hours

C short day species with a critical photoperiod of 8 hours

D short day species with a critical photoperiod of 14 hours.

**24.** The following list shows the effects of drugs on fetal development in humans.

1 Limb deformation

2 Overall reduction of growth

3 Slowing of mental development

Which line in the table below correctly matches alcohol, nicotine and thalidomide with their effects on fetal development?

| | *Alcohol* | *Nicotine* | *Thalidomide* |
|---|---|---|---|
| A | 2 only | 2 and 3 only | 1 and 3 only |
| B | 3 only | 2 and 3 only | 1 only |
| C | 2 and 3 only | 2 only | 1 and 3 only |
| D | 2 and 3 only | 2 and 3 only | 1 only |

**25.** Which line in the table below identifies correctly the hormones which stimulate the conversion of glucose and glycogen?

| | *glycogen → glucose* | *glucose → glycogen* |
|---|---|---|
| A | glucagon and adrenalin | insulin |
| B | adrenalin | glucagon and insulin |
| C | insulin | adrenalin and glucagon |
| D | glucagon and insulin | adrenalin |

**26.** Drinking a large volume of water will lead to

A increased production of ADH and kidney tubules becoming more permeable to water

B decreased production of ADH and kidney tubules becoming less permeable to water

C increased production of ADH and kidney tubules becoming less permeable to water

D decreased production of ADH and kidney tubules becoming more permeable to water.

**[Turn over**

**27.** The graph below shows how the concentration of insulin in the blood varies with the concentration of glucose in the blood.

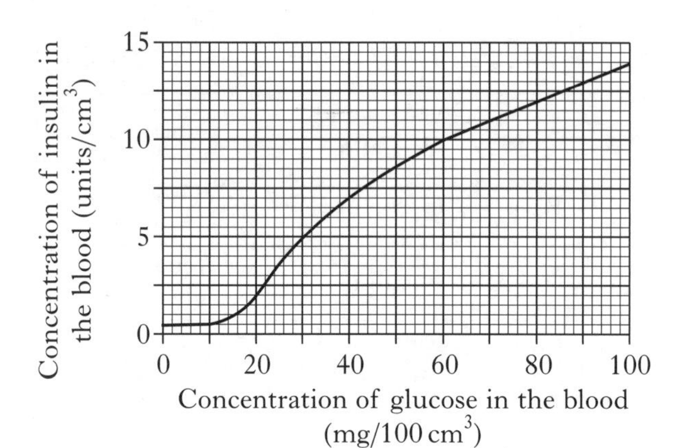


What total mass of glucose would be present at an insulin concentration of 10 units/$cm^3$, in an individual with 5 litres of blood?

A 60 mg

B 300 mg

C 3000 mg

D 6000 mg

**28.** The graph below shows the annual variation in the biomass and population density of *Corophium*, a small burrowing invertebrate found in the mud of most Scottish estuaries.

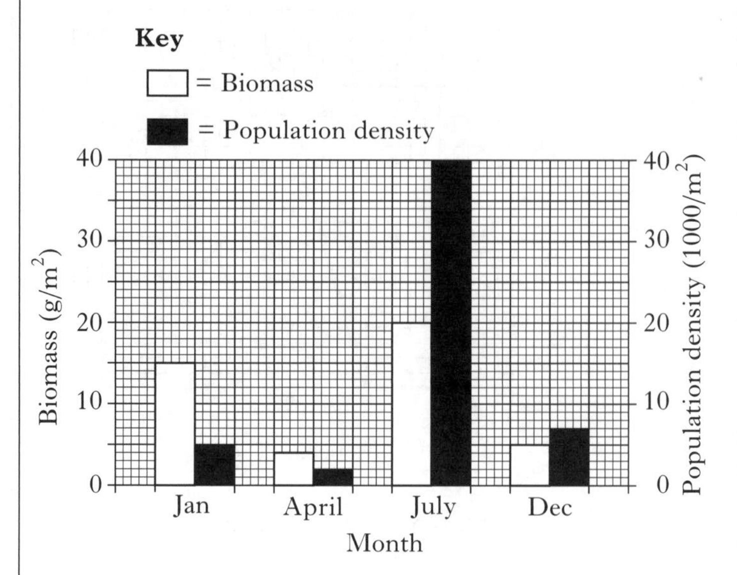


During which month do individual *Corophium* have the greatest average mass?

A January

B April

C July

D December

**29.** The graph below shows the effect of air temperature on the metabolic rate of two different animals.

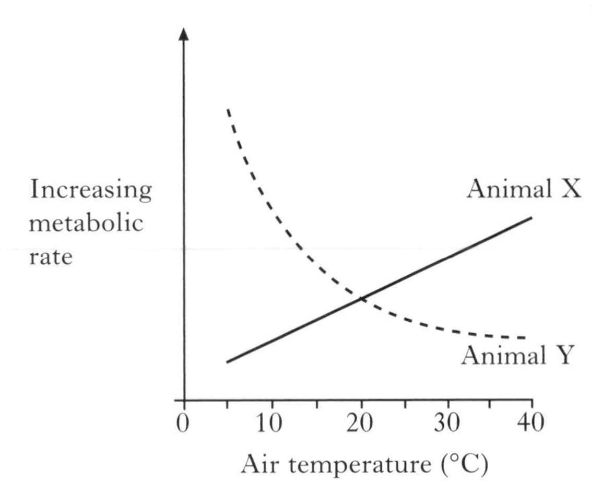


Which line in the table below identifies correctly the temperatures at which oxygen consumption will be greatest in the tissues of each animal?

| | *Animal X* | *Animal Y* |
|---|---|---|
| A | 20 °C | 20 °C |
| B | 40 °C | 40 °C |
| C | 40 °C | 5 °C |
| D | 5 °C | 40 °C |

**30.** Which of the following comparisons of early and late succession in plant communities in their habitat is correct?

| | *Early succession* | *Late succession* |
|---|---|---|
| A | low biomass | high biomass |
| B | complex food webs | simple food webs |
| C | soil is deep | soil is shallow |
| D | high species diversity | low species diversity |

**Candidates are reminded that the answer sheet MUST be returned INSIDE the front cover of this answer book.**

**[Turn over**

DO NOT WRITE IN THIS MARGIN

*Marks*

**2.** (*a*) State the **exact** location of photosynthetic pigments in plant leaf cells.

_______________________________________________ **1**

(*b*) The table below shows the mass of photosynthetic pigments in the leaves of two plant species.

| *Photosynthetic pigment* | *Mass of photosynthetic pigment in the leaves* (μg per $cm^3$ of leaf) | |
|---|---|---|
| | *Species A* | *Species B* |
| chlorophyll a | 0·92 | 0·93 |
| chlorophyll b | 0·34 | 0·35 |
| carotene | 0·32 | 0·65 |
| xanthophyll | 0·28 | 0·55 |

Which species is best adapted to grow in the shade of taller plants?

Explain your choice.

Species ________________

Explanation _______________________________________________

_______________________________________________

_______________________________________________ **1**

DO NOT WRITE IN THIS MARGIN

*Marks*

**2. (continued)**

(*c*) The diagram below shows some events in the carbon fixation stage (Calvin cycle) of photosynthesis in a plant kept in bright light.

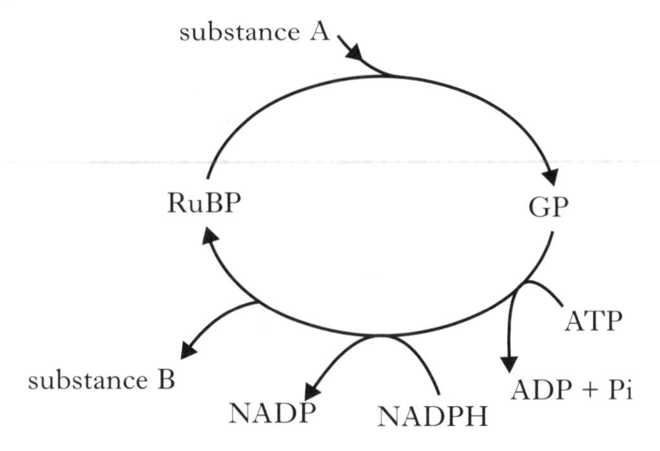


(i) Name substances A and B.

A ______________________________

B ______________________________ **2**

(ii) NADP carries hydrogen to the carbon fixation stage.

Describe the role of hydrogen in the carbon fixation stage.

________________________________________________

________________________________________________ **1**

(iii) Complete the table below to show the number of carbon atoms in one molecule of each compound.

| *Compound* | *Number of carbon atoms per molecule* |
|---|---|
| RuBP | |
| GP | |

**1**

(iv) Predict what would happen to the concentrations of RuBP and GP in leaf cells if the plant was moved from bright light into dark conditions.

Explain your answer.

RuBP ______________________________________________

GP ________________________________________________ **1**

Explanation ________________________________________

________________________________________________ **1**

DO NOT WRITE IN THIS MARGIN

*Marks*

**6.** Honey bees are social insects. They forage for food at various distances from their hive. When a bee finds a food source, it returns to the hive and communicates the location of the food to other bees using body movements called waggle dances. These are performed several times with short intervals between them.

In an investigation, bees were fitted with radio tracking devices which allowed the distances they travelled from the hive to be measured. The waggle dances performed by each returning bee were studied. The average time taken for its waggle dance was recorded and the number of times it was performed in 15 seconds was counted.

The results are shown in the graph below.

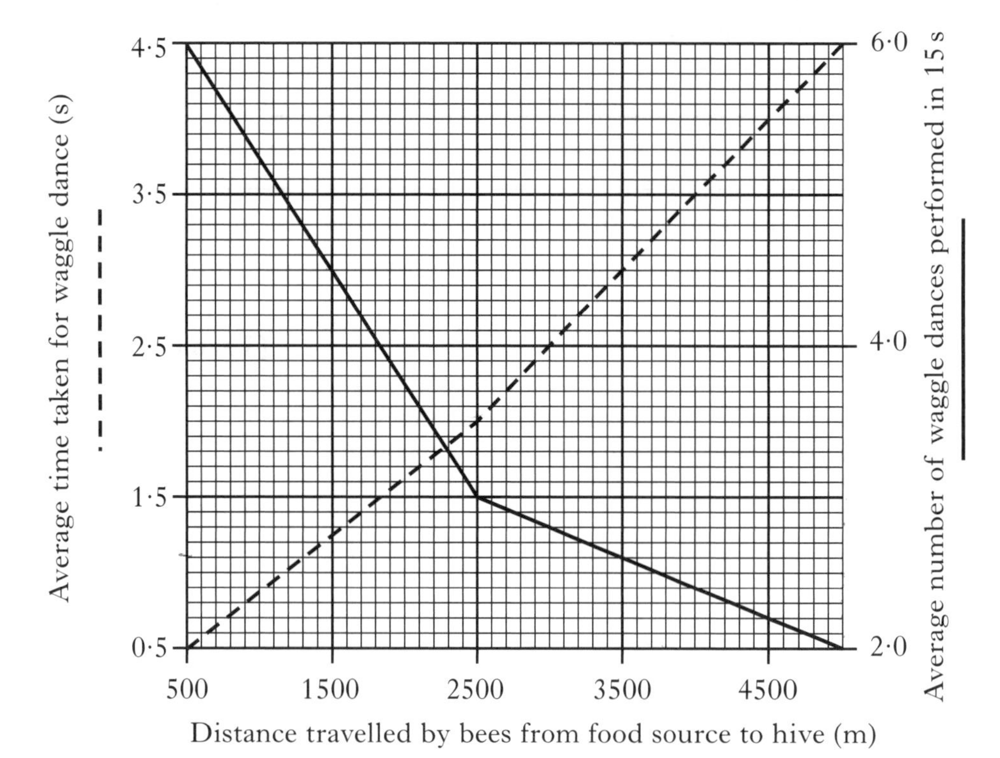


(*a*) (i) **Use values from the graph** to describe the relationship between the distance travelled by a bee from the food source and the number of waggle dances performed in 15 s.

______________________________________________

______________________________________________

______________________________________________ **2**

(ii) State the average time taken for a waggle dance when the distance travelled by a bee from the food source is 1500 m.

______________________ s **1**

DO NOT WRITE IN THIS MARGIN

*Marks*

**6. (a) (continued)**

(iii) Calculate the percentage increase in the average time taken for a waggle dance when the distance from the food source to the hive increases from 500 to 3500 metres.

*Space for calculation*

____________% **1**

(iv) Predict the total time a bee would spend in the waggle dances in a 15 s period when the food source is 2500 m away from the hive.

____________ s **1**

(b) In another investigation, the waggle dances of six bees from another hive were observed.

The results are shown in the table below.

| BEE | *Number of times waggle dance was performed in 15 s* |
|---|---|
| 1 | 2·65 |
| 2 | 2·20 |
| 3 | 2·30 |
| 4 | 2·55 |
| 5 | 2·70 |
| 6 | 2·60 |
| Average | |

(i) Complete the table by calculating the average number of times the waggle dance was performed in 15 s.

*Space for calculation*

**1**

(ii) Using information from the table and the graph, predict the distance that was travelled by bee 6 to its food source.

____________ m **1**

(c) (i) Apart from its distance from the hive, what other information about food sources would be useful to the bees?

________________________________________ **1**

(ii) In terms of the economics of foraging, explain the advantage of waggle dances to bees.

________________________________________

________________________________________ **1**

DO NOT WRITE IN THIS MARGIN

*Marks*

**7.** The diagram below shows a pair of homologous chromosomes during meiosis. P and Q show points where crossing over **may** occur. The other letters show the positions of the alleles of four genes.

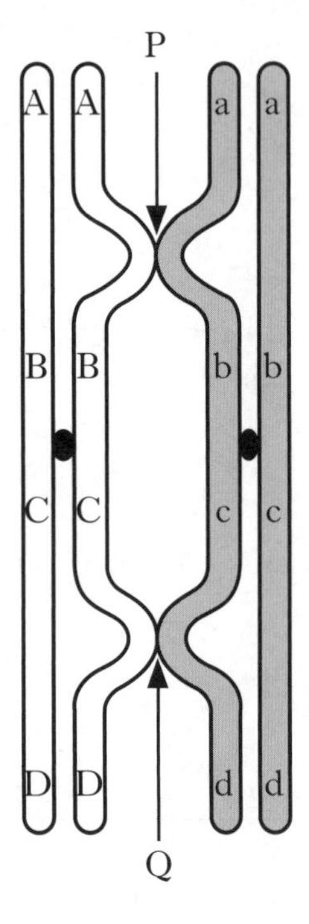


(*a*) What evidence confirms that these chromosomes are homologous?

______________________________________________

______________________________________________ **1**

(*b*) (i) What name is given to points P and Q?

______________________________ **1**

(ii) State the importance of crossing over in meiosis.

______________________________________________

______________________________________________ **1**

DO NOT WRITE IN THIS MARGIN

*Marks*

**7. (continued)**

(*c*) (i) In the table below, tick (✓) the boxes to identify which combination of alleles would result from crossing over at point P only or crossing over at both points P and Q on the diagram.

| | *Crossing over at* | |
|---|---|---|
| *Combination of alleles* | *point P only* | *both points P and Q* |
| Abcd | | |
| aBCD | | |
| AbcD | | |
| aBCd | | |

**1**

(ii) Give **one** possible sequence of alleles which could be found in a recombinant gamete formed if crossing over occurred at point Q only.

______________________________ **1**

**[Turn over**

DO NOT WRITE IN THIS MARGIN

Marks

**13.** The diagrams show three barley seedlings which were grown in culture solutions. Solution A had all elements required for plant growth. Solutions B and C were each missing in one element required for normal plant growth. The seedlings were kept under a lamp which provided constant bright light conditions.

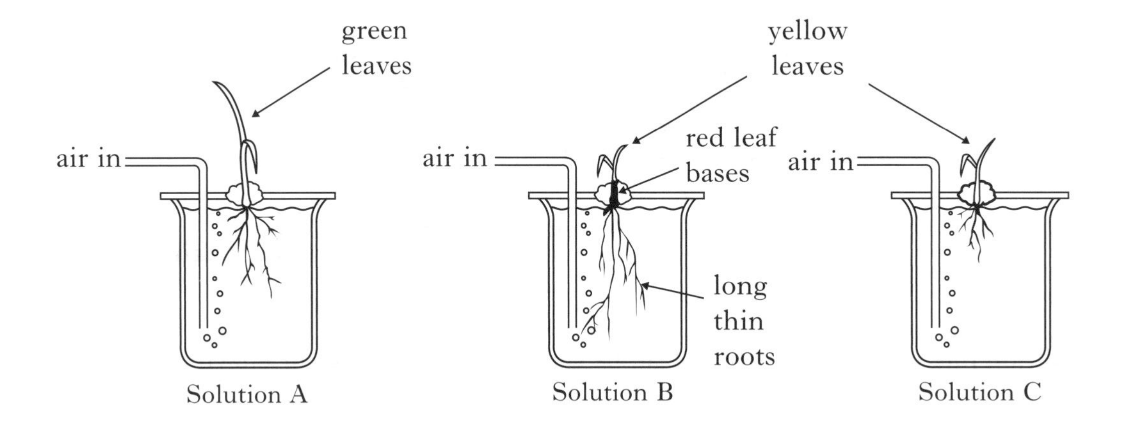


(*a*) Complete the table below by naming the element that was missing from culture solutions B and C and by giving a role for the element missing from solution B.

| *Solution* | *Element missing from solution* | *Role of element in plants* |
|---|---|---|
| B | | |
| C | | component of chlorophyll |

2

(*b*) In a further experiment, the seedling in solution A was grown in complete darkness for a week.

Give the term which would describe the seedling after this period and describe how the treatment would have affected its appearance.

Term ______________________________ 1

Description ______________________________

______________________________ 1

DO NOT WRITE IN THIS MARGIN

*Marks*

**14.** (*a*) The Colorado beetle is a pest of potato crops. In an investigation, the population of beetles in a 2000 $m^2$ potato field was estimated as described below.

A sample of the beetles from the field was collected and counted.

Each beetle was marked with a spot of paint then released back into the field.

Three days later a second sample of beetles was collected and counted.

The number of marked beetles in this second sample was noted.

The results are shown in the table below.

| *Number of beetles that were marked and released* | *Number of beetles in second sample* | *Number of marked beetles in second sample* |
|---|---|---|
| 500 | 450 | 5 |

The population of beetles can be estimated using the following formula.

$$\text{Population} = \frac{\text{number of beetles marked and released} \times \text{number of beetles in second sample}}{\text{number of marked beetles in second sample}}$$

(i) Calculate the **population density** of beetles in the field.

*Space for calculation*

__________ beetles per $m^2$ **1**

(ii) The beetle population is affected by both density-dependent and density-independent factors.

Name a density-dependent and a density-independent factor that could affect the population of beetles in the field.

Density-dependent factor ____________________

Density-independent factor ____________________ **1**

(*b*) A population of a species may be monitored to gain data for use in pest control.

State **two** further reasons why a wild population may be monitored.

1 ____________________

2 ____________________ **1**

**[Turn over for Section C on *Page thirty-two***

DO NOT WRITE IN THIS MARGIN

*Marks*

## SECTION C

**Both questions in this section should be attempted.**

Note that each question contains a choice.

**Questions 1 and 2 should be attempted on the blank pages which follow.**

**Supplementary sheets, if required, may be obtained from the Invigilator.**

**All answers must be written clearly and legibly in ink.**

**Labelled diagrams may be used where appropriate.**

**1.** Answer **either** A **or** B.

**A.** Write notes on plant growth and development under the following headings:

(i) the effects of indole acetic acid (IAA); **6**

(ii) the role of gibberellic acid (GA) in the germination of barley grains. **4**

**(10)**

**OR**

**B.** Write notes on the following:

(i) endotherms and ectotherms; **2**

(ii) temperature regulation in mammals. **8**

**(10)**

**In question 2, ONE mark is available for coherence and ONE mark is available for relevance.**

**2.** Answer **either** A **or** B.

**A.** Give an account of the importance of isolating mechanisms, mutations and natural selection in the evolution of new species. **(10)**

**OR**

**B.** Give an account of the transpiration stream and its importance to plants. **(10)**

[*END OF QUESTION PAPER*]

DO NOT WRITE IN THIS MARGIN

**SPACE FOR ANSWERS**

DO NOT WRITE IN THIS MARGIN

**SPACE FOR ANSWERS**

DO NOT WRITE IN THIS MARGIN

**SPACE FOR ANSWERS**

DO NOT WRITE IN THIS MARGIN

**SPACE FOR ANSWERS**

ADDITIONAL GRAPH PAPER FOR QUESTION 10(*b*)

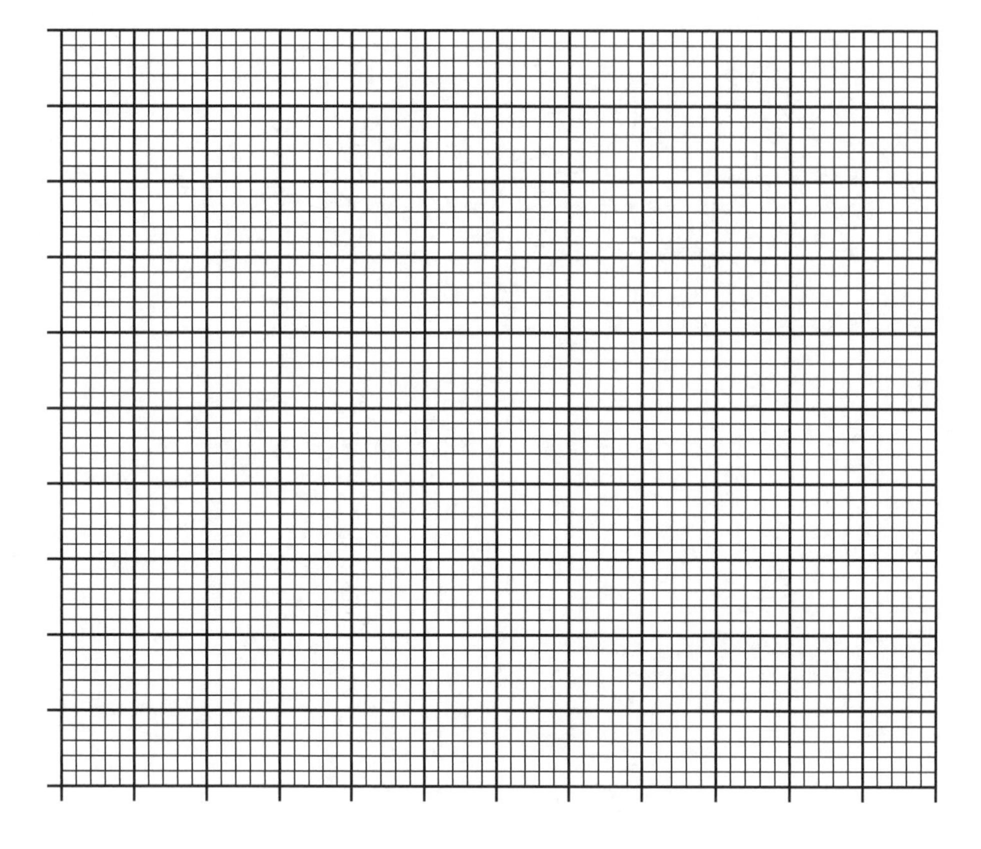

HIGHER

2011

**3.** The diagram below represents a summary of respiration in a mammalian muscle cell.

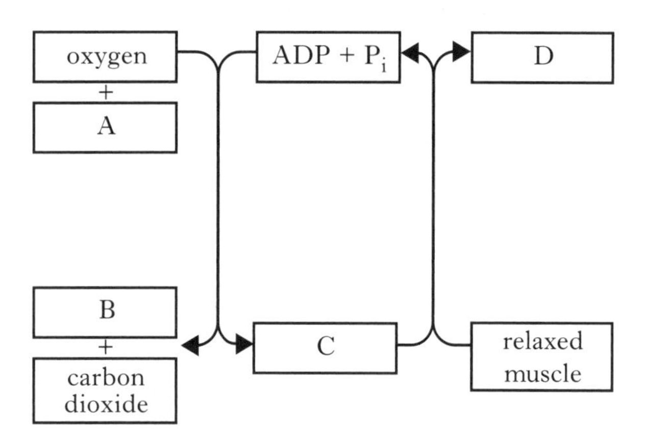


Which box represents ATP?

**4.** Which of the following produces water?

A Krebs cycle

B Glycolysis

C Photolysis

D Cytochrome system

**5.** The graph below shows changes which occur in the masses of protein, fat and carbohydrate in the body of a hibernating mammal during seven weeks without food.

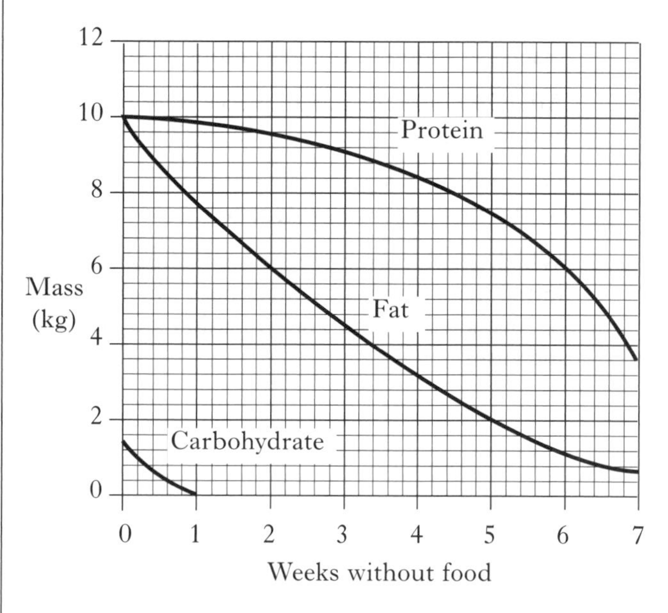


What percentage of the original mass of fat was used up between weeks 2 and 5?

A 33%

B 40%

C 67%

D 80%

**6.** Which of the following compounds are linked by peptide bonds to form more complex molecules?

A Bases

B Nucleic acids

C Nucleotides

D Amino acids

**7.** A DNA molecule consists of 4000 nucleotides, of which 20% contain the base adenine.

How many of the nucleotides in this DNA molecule will contain guanine?

A 800

B 1000

C 1200

D 1600

**8.** The diagram below shows parts of an animal cell.

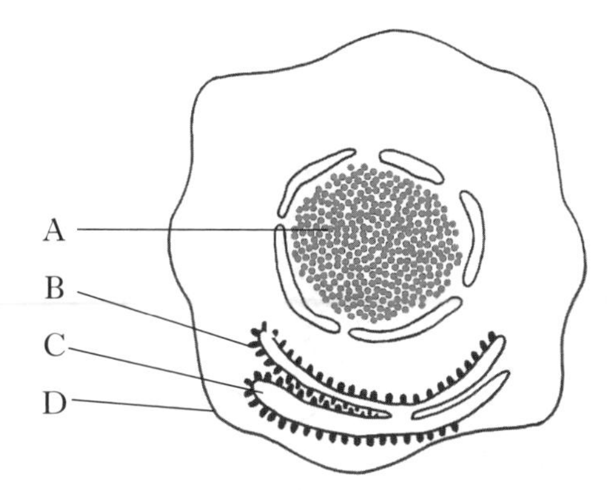


Where does synthesis of mRNA take place?

**9.** The function of tRNA in cell metabolism is to

A transport amino acids to be used in synthesis

B carry codons to the ribosomes

C synthesise proteins

D transcribe the DNA code.

**10.** Which of the following describes a cellular defence mechanism in plants?

A Growth of sharp spines

B Production of cellulose fibres

C Development of low meristems

D Secretion of sticky resin

**11.** Huntington's Disease is an inherited condition in humans caused by a dominant allele. A woman's father is heterozygous for the condition. Her mother is not affected by the condition.

What is the chance of the woman being affected by the condition?

A 1 in 1

B 1 in 2

C 1 in 3

D 1 in 4

**12.** In guinea pigs, brown hair **B** is dominant to white hair **b** and short hair **S** is dominant to long hair **s**.

A brown, long-haired male was crossed with a white, short-haired female. The $F_1$ phenotype ratio was

1 brown, short-haired:
1 white, short-haired:
1 brown, long-haired:
1 white, long-haired.

What were the genotypes of the parents?

| | *Male* | *Female* |
|---|---|---|
| A | BbSs | BbSs |
| B | Bbss | bbSs |
| C | BBss | bbSS |
| D | bbSs | Bbss |

**13.** The following cross was carried out using two true-breeding strains of the fruit fly, *Drosophila*.

Parents: straight wing black body × curly wing grey body

$F_1$: all straight wing black body

$F_1$ allowed to interbreed

$F_2$: 3 straight wing black body : 1 curly wing grey body

The result would suggest that

A crossing over has occurred between the genes

B before isolation, $F_1$ females had mated with their own-type males

C non-disjunction of chromosomes in the sex cells has taken place

D these genes are linked.

[Turn over

**18.** The diagram below shows a cross section through the stem of a hydrophyte.

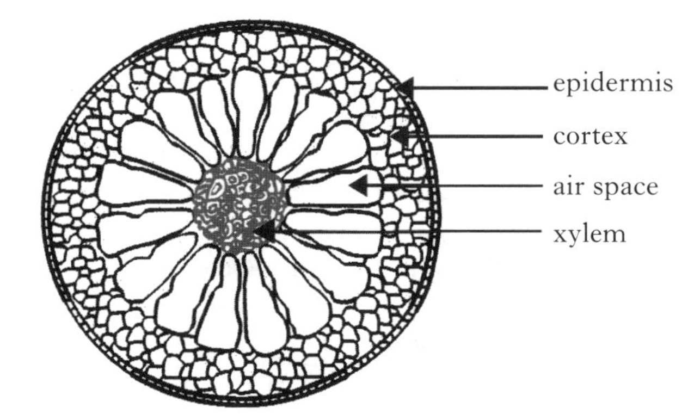


The arrangement of the xylem is of benefit to the plant because it

A gives the stem flexibility in flowing water

B allows uptake of water through the cortex

C gives the stem increased support

D allows transport of sugars to the roots.

**19.** The statements below relate to bird behaviour.

1 Blackbirds sing to mark their territory.

2 Arctic and common terns form large mixed breeding colonies.

3 Black grouse gather on open areas of short grass and males display to females.

4 Great skuas chase other seabirds and force them to drop their food.

Which of the above statements are related to intraspecific competition?

A 1 and 2 only

B 1 and 3 only

C 2 and 4 only

D 3 and 4 only

**20.** When the intensity of grazing by herbivores increases in a grassland ecosystem, diversity of plant species may increase as a result.

Which statement explains this observation?

A Few herbivores are able to eat every plant species present.

B Grazing stimulates growth in some plant species.

C Vigorous plant species are eaten so less competitive species can now thrive.

D Plant species with defences against herbivores are selected.

**21.** The diagram below shows a section through a woody twig.

Which label shows the position of a meristem?

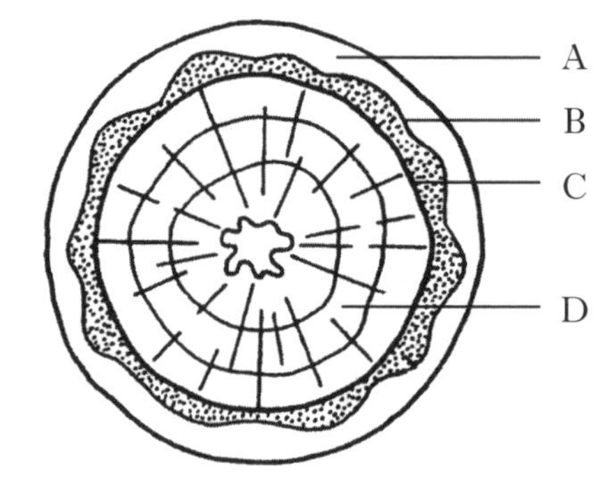


**22.** In the condition phenylketonuria (PKU), the human body is unable to

A synthesise phenylalanine from tyrosine

B secrete phenylalanine from cells

C absorb phenylalanine into the bloodstream

D convert phenylalanine to tyrosine.

**23.** The diagram below shows an experiment to investigate the role of IAA in the growth of lateral buds.

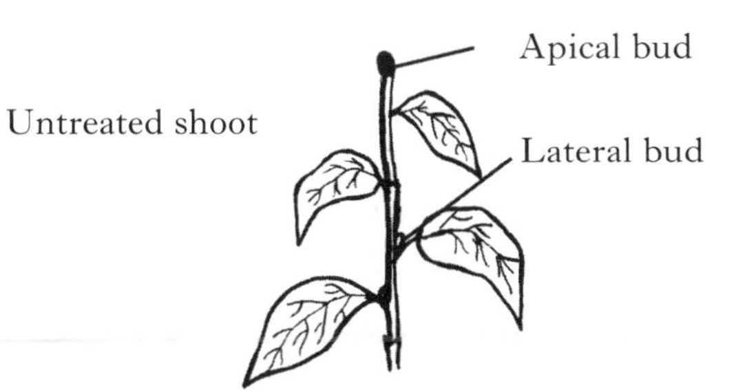


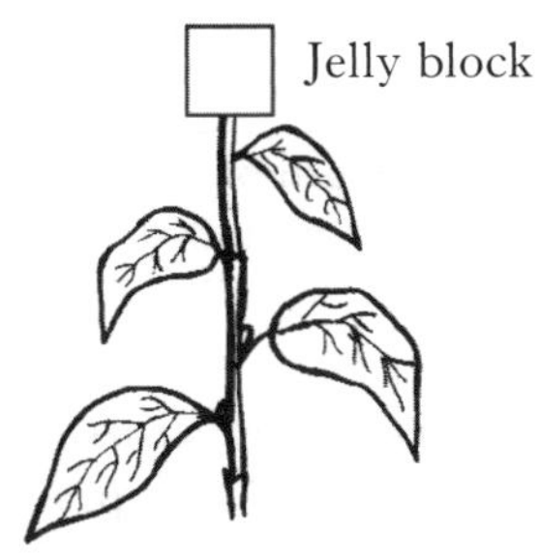


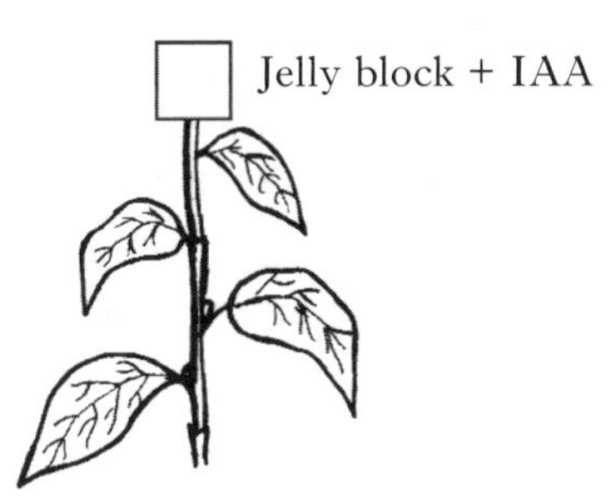


Which line in the table correctly shows the expected growth of lateral buds in the experiment?

✓ = growth of lateral buds
✗ = no growth of lateral buds

| | *Untreated shoot* | *Apical bud removed and replaced by jelly block* | *Apical bud removed and replaced by a jelly block containing IAA* |
|---|---|---|---|
| A | ✗ | ✓ | ✗ |
| B | ✗ | ✗ | ✓ |
| C | ✓ | ✓ | ✗ |
| D | ✓ | ✗ | ✓ |

**24.** The table below shows the results of an experiment to investigate the effect of IAA on the development of roots from sections of pea stems.

| *Concentration of IAA* (units) | *Average number of roots per stem section* |
|---|---|
| 2 | 2·0 |
| 4 | 2·2 |
| 6 | 3·8 |
| 8 | 5·7 |
| 10 | 6·6 |

The greatest percentage increase in the average number of roots per stem section is caused by an increase in IAA concentration (units) from

A 2 to 4

B 4 to 6

C 6 to 8

D 8 to 10.

**[Turn over**

DO NOT WRITE IN THIS MARGIN

Marks

**SECTION B**

**All questions in this section should be attempted.**

**All answers must be written clearly and legibly in ink.**

**1.** The diagram below shows *Paramecium*, a unicellular organism found in fresh water.

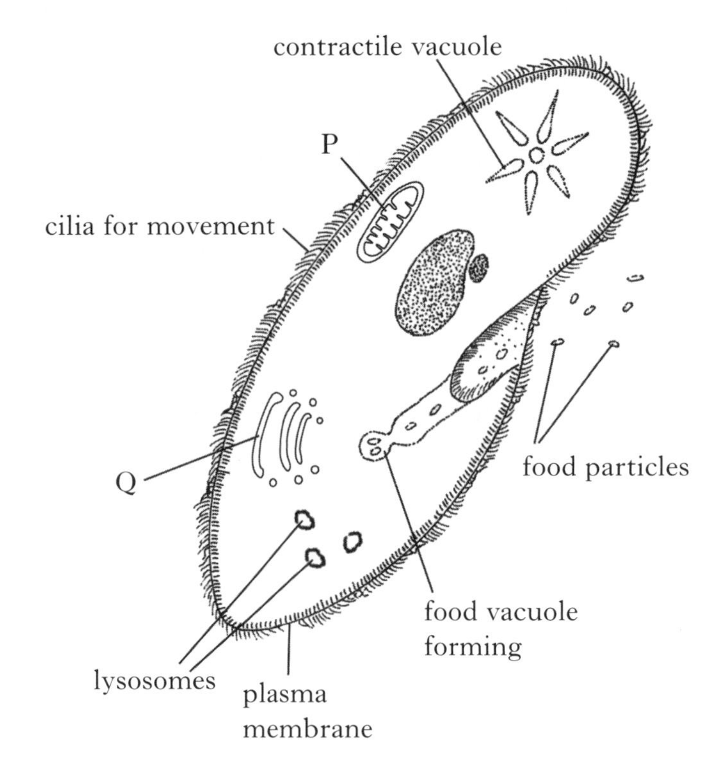


(*a*) Identify organelles P and Q.

P ______________________

Q ______________________ **2**

(*b*) (i) Name **two** chemical components of the plasma membrane.

1 ______________________________

2 ______________________________ **1**

(ii) Give a property of the plasma membrane which is related to its role in osmosis.

______________________________________________ **1**

*Marks*

**1. (continued)**

(*c*) *Paramecium* has contractile vacuoles that fill with excess water which has entered the organism by osmosis. These vacuoles contract to remove this water from the organism.

The rate of contraction of the vacuoles is affected by the concentration of the solution in which the *Paramecium* is found.

In which solution would the highest rate of contraction of the vacuoles occur?

Underline the correct answer.

**hypertonic** **hypotonic** **isotonic** 1

(*d*) *Paramecium* feeds on micro-organisms present in water.

Use information from the diagram to describe how *Paramecium* obtains and digests food.

________________________________________

________________________________________

________________________________________ 2

**[Turn over**

DO NOT WRITE IN THIS MARGIN

*Marks*

**2.** The diagram below shows an outline of respiration in yeast cells.

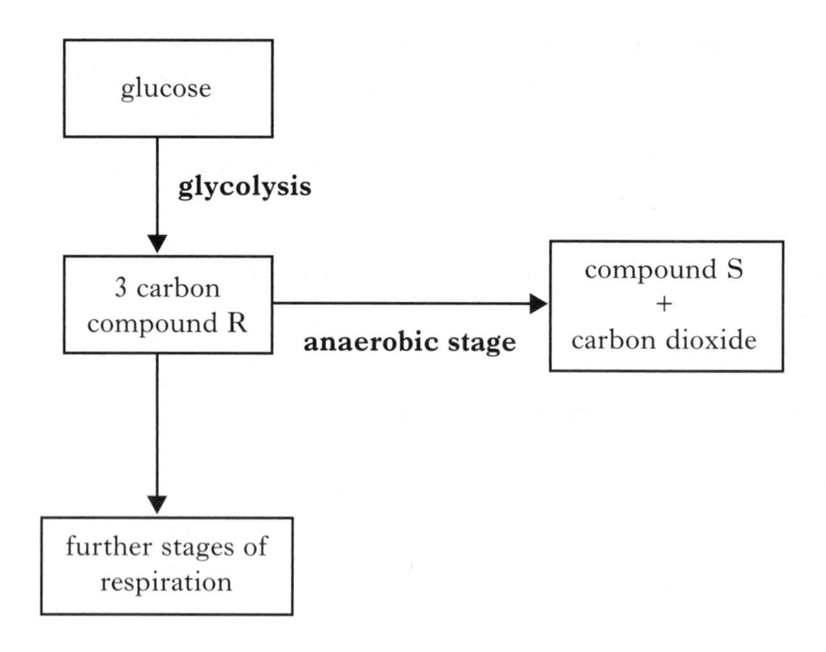


(*a*) State the location of glycolysis in yeast cells.

_______________________________ **1**

(*b*) Name **one** substance, other than glucose, which must be present for glycolysis to occur.

_______________________________ **1**

(*c*) Name compounds R and S.

R _______________________________ **1**

S _______________________________ **1**

(*d*) Explain why the further stages of respiration cannot occur in anaerobic conditions.

_______________________________

_______________________________

_______________________________ **1**

DO NOT WRITE IN THIS MARGIN

*Marks*

**3.** The graph shows the rate of potassium ion uptake by human liver cells in different oxygen concentrations at 30 °C.

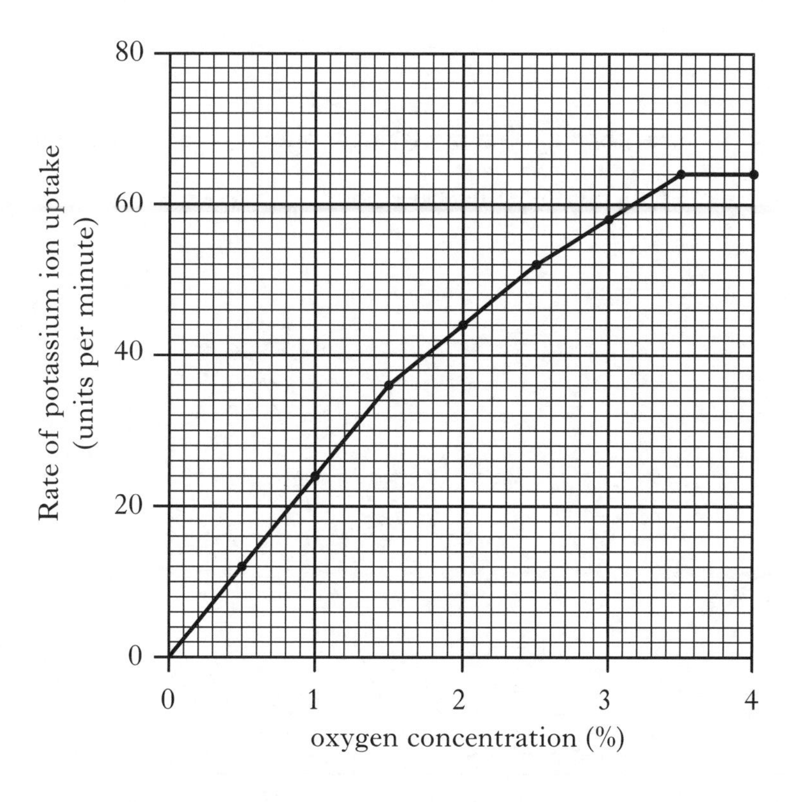


(*a*) When the oxygen concentration is 1%, how many units of potassium would a cell take up in one hour?

*Space for calculation*

________________ units per hour **1**

(*b*) Suggest a reason why the graph levels off at oxygen concentrations above 3·5%.

____________________________________________

____________________________________________ **1**

(*c*) When the experiment was repeated at 20 °C, the potassium ion uptake decreased. Explain this observation.

____________________________________________

____________________________________________

____________________________________________

____________________________________________ **2**

DO NOT WRITE IN THIS MARGIN

*Marks*

**4.** (*a*) The diagram below shows the absorption spectrum of a single photosynthetic pigment from a plant and the rate of photosynthesis of the plant in different colours of light.

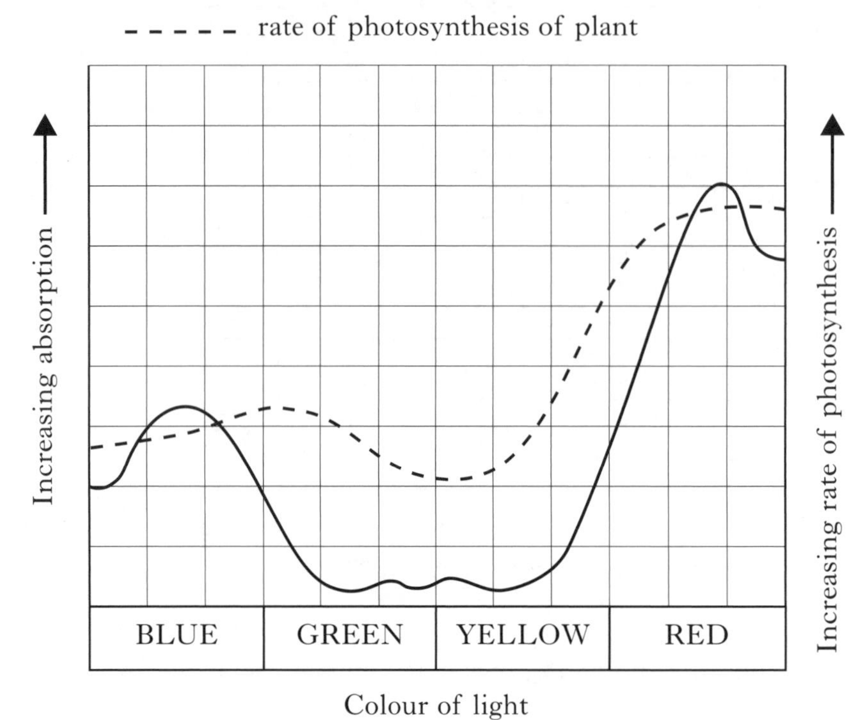


(i) Leaves of this plant contain more than one photosynthetic pigment.

Use evidence from the graph to justify this statement.

______________________________________________

______________________________________________

______________________________________________ **1**

(ii) Name a technique used to separate mixtures of photosynthetic pigments.

______________________________ **1**

DO NOT WRITE IN THIS MARGIN

Marks

**4. (continued)**

(*b*) *Spirogyra* is a photosynthetic green alga which grows as a long strand of cells. A strand of *Spirogyra* was placed into water containing aerobic bacteria. Different parts of the strand were exposed to different colours of light. After a period of time, the bacteria had moved into the positions shown in the diagram below.

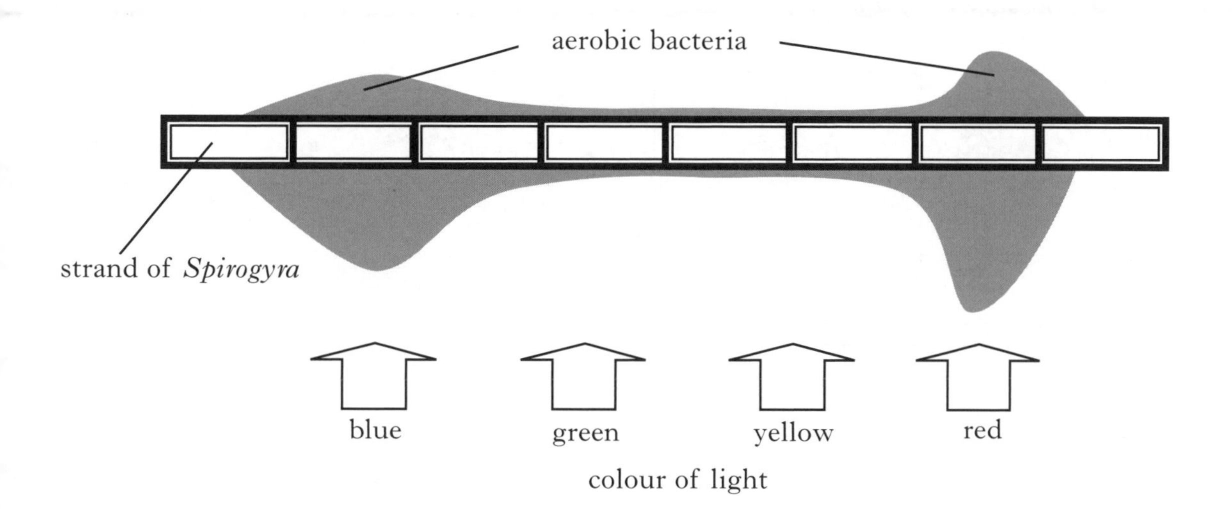


Explain the distribution of aerobic bacteria shown in the diagram.

______________________________________________

______________________________________________

______________________________________________

______________________________________________ **2**

**[Turn over**

DO NOT WRITE IN THIS MARGIN

*Marks*

**5.** (*a*) Eye colour in fruit flies is sex-linked.

Red eye colour **R** is dominant to white eye colour **r**.

A heterozygous red-eyed female fly was crossed with a white-eyed male.

(i) Complete the grid by adding the genotypes of

1 the male and female gametes; **1**

2 the possible offspring. **1**

| | | *Female gametes* | |
|---|---|---|---|
| | | | |
| *Male gametes* | | | |
| | | | |

(ii) Tick (✓) the box(es) to show all the expected phenotypes of the offspring from this cross.

red-eyed female ☐ white-eyed female ☐

red-eyed male ☐ white-eyed male ☐ **1**

(iii) Explain why the actual phenotype **ratio** obtained from this cross could differ from the expected.

______________________________________________

______________________________________________ **1**

DO NOT WRITE IN THIS MARGIN

*Marks*

**5. (continued)**

(*b*) Genes K, L, M and N are located on the same chromosome in fruit flies.

The recombination frequencies of pairs of these genes are given in the table.

| *Genes* | *Recombination frequency (%)* |
|---|---|
| K and L | 18 |
| N and L | 25 |
| M and N | 17 |
| L and M | 8 |
| K and N | 7 |

Complete the diagram below to show the relative positions of genes L, M and N on the chromosome.

K

**1**

**[Turn over**

DO NOT WRITE IN THIS MARGIN

*Marks*

**6.** The compensation point is the light intensity at which a plant's carbon dioxide uptake by photosynthesis is equal to its carbon dioxide output from respiration.

In some plant species, compensation point can be reduced when the plant is moved from bright light to shaded conditions.

**Graph 1** shows how the compensation points of three species of plant changed over a 25 day period after they were moved from bright light into shaded conditions.

**Graph 1**

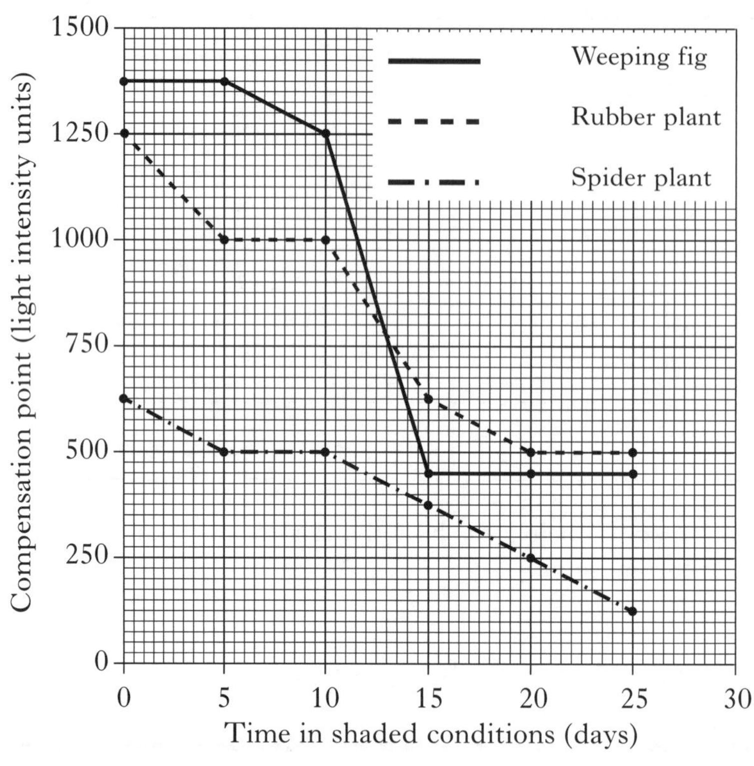


(*a*) (i) Use values from **Graph 1** to describe the changes in compensation point of the weeping fig over the 25 day period.

**2**

(ii) Calculate the percentage decrease in compensation point of the rubber plant over the 25 day period.

*Space for calculation*

______ % **1**

(iii) Predict the compensation point of the spider plant at **28 days**.

______ light intensity units **1**

(iv) Use evidence from the graph to explain why the rubber plant could not grow successfully in a constant light intensity of 400 light intensity units.

**2**

DO NOT WRITE IN THIS MARGIN

*Marks*

**6. (continued)**

(*b*) After 10 days in shaded conditions, a plant of one species was placed into different light intensities and its carbon dioxide output and uptake were measured. This was repeated with another plant of the **same** species which had been in the shade for 20 days.

The results are shown in **Graph 2**.

**Graph 2**

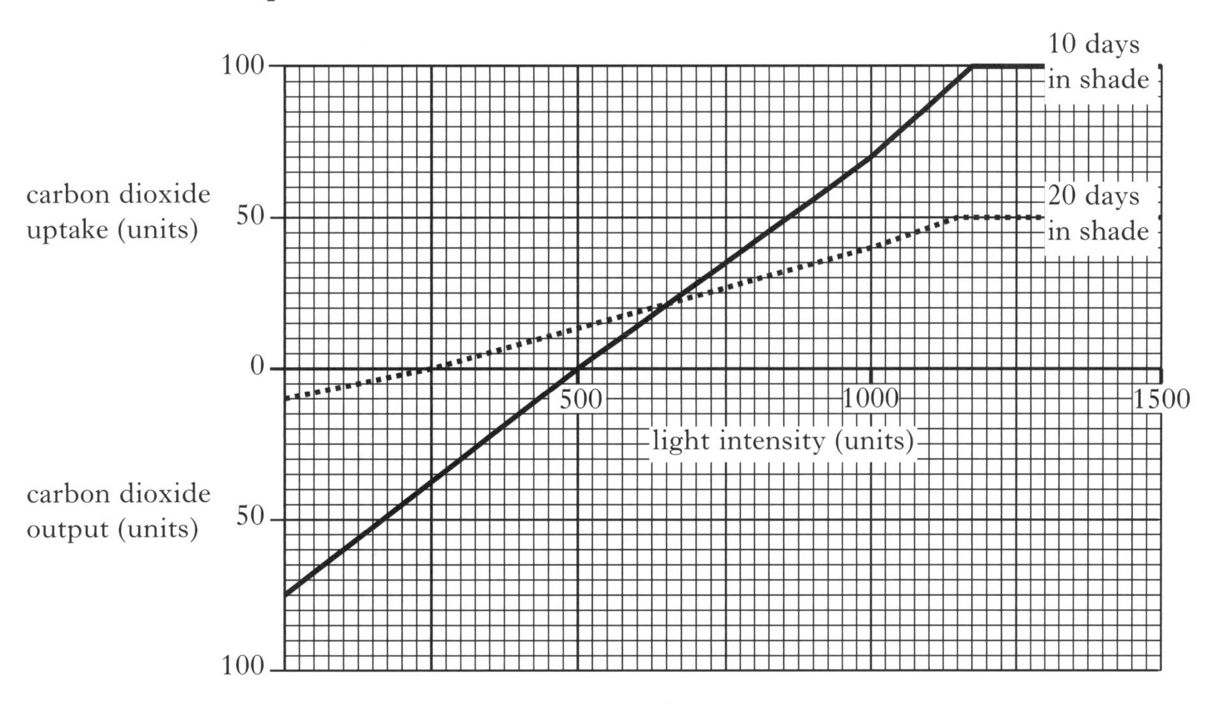


(i) Use all the data to identify the plant species referred to in **Graph 2**.

Tick (✓) the correct box and give a reason for your choice.

weeping fig ☐ rubber plant ☐ spider plant ☐

Reason

________________________________________

________________________________________ **1**

(ii) From **Graph 2**, calculate how many times greater the carbon dioxide uptake of this plant was at 1000 units of light intensity after 10 days in shaded conditions compared with after 20 days in shaded conditions.

*Space for calculation*

__________ times **1**

DO NOT WRITE IN THIS MARGIN

*Marks*

**9.** (*a*) The table shows behavioural adaptations of lions for obtaining food.

(i) Complete the table below by explaining how each adaptation is beneficial for obtaining food.

| *Behavioural adaptation* | *Benefit for obtaining food* |
|---|---|
| Cooperative hunting | |
| Territorial behaviour | |

**2**

(ii) These behavioural adaptations ensure that lions can forage economically.

Explain what is meant by this statement in terms of energy gained and lost.

______________________________

______________________________

______________________________ **1**

(iii) Lions hunt wildebeest, which live in large herds.

Explain how living in large herds benefits the wildebeest in terms of predation by lions.

______________________________

______________________________

______________________________ **1**

DO NOT WRITE IN THIS MARGIN

*Marks*

**9. (continued)**

(*b*) If a snail is disturbed, it withdraws into its shell and re-emerges a few minutes later.

(i) Name the type of behaviour shown by the withdrawal response.

______________________________ **1**

(ii) What is the advantage to a snail of withdrawing into its shell?

______________________________

______________________________ **1**

**[Turn over**

**10.** Gibberellic acid (GA) is needed to break dormancy of rice grains allowing them to germinate.

An experiment was carried out to investigate the effects of GA on the germination of rice grains.

30 $cm^3$ of different concentrations of GA solution was placed into separate beakers. 50 rice grains were added to each beaker. Each beaker was then covered with plastic film.

After 12 hours, the grains were removed from the solutions and evenly spaced in separate dishes on filter paper soaked with 20 $cm^3$ of water.

The dishes were covered and kept in the dark for 10 days and the number of germinated grains in each dish was counted.

A second batch of grains was treated in the same way but these were left in the GA solutions for 36 hours.

The results are shown in the table.

| *Concentration of GA solution (mg per litre)* | *Number of rice grains germinated* | |
|---|---|---|
| | *After 12 hours in GA solution* | *After 36 hours in GA solution* |
| 0 | 5 | 6 |
| 5 | 7 | 14 |
| 10 | 16 | 31 |
| 20 | 23 | 35 |
| 30 | 28 | 41 |
| 60 | 31 | 43 |

DO NOT WRITE IN THIS MARGIN

*Marks*

(*a*) Identify **one** variable, not already described, that should be kept constant.

______________________________ **1**

(*b*) (i) Explain how the solution with 0 mg per litre GA acts as a control in this experiment.

______________________________

______________________________ **1**

(ii) Suggest why some germination occurs in the control.

______________________________ **1**

(*c*) Identify a feature of the experimental procedure which ensured the reliability of the results.

______________________________ **1**

DO NOT WRITE IN THIS MARGIN

*Marks*

**10. (continued)**

(*d*) Predict how the concentration of GA in the beakers would have been affected if they had not been covered with plastic film.

Underline the correct answer and give a reason for your choice.

**increased** **decreased** **stayed the same**

Reason ______________________________

______________________________ 1

(*e*) Calculate the difference in **percentage** germination between the grains kept in the 5 mg GA per litre solution for 12 hours and those kept in the 30 mg GA per litre solution for 12 hours.

*Space for calculation*

______________ % 1

(*f*) On the grid below, draw a line graph to show the number of grains germinated after 36 hours in the different concentration of GA solution.

(Additional graph paper, if required, will be found on *Page forty*.)

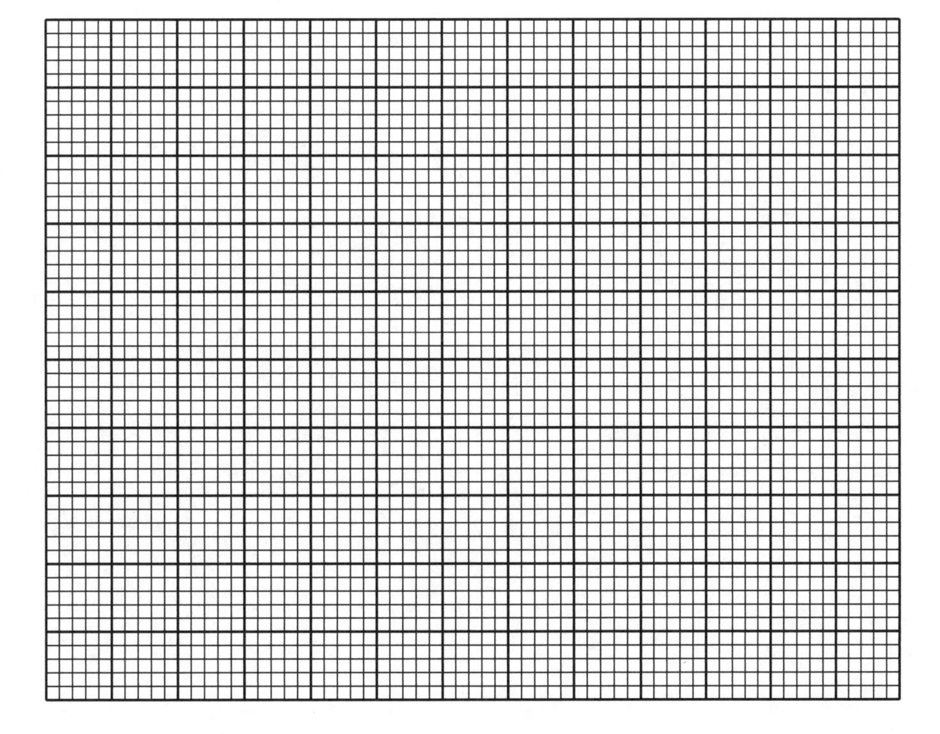

2

(*g*) Give **two** conclusions which can be drawn from the results in the **table**.

1 ______________________________

2 ______________________________ 2

DO NOT WRITE IN THIS MARGIN

*Marks*

**10. (continued)**

(*h*) GA induces the release of amylase in the germinating grains of plants such as rice and barley.

Name the site within the grains which produces amylase.

________________________________________ **1**

DO NOT WRITE IN THIS MARGIN

*Marks*

**11.** The graph below shows how the body length of a locust changes over time.

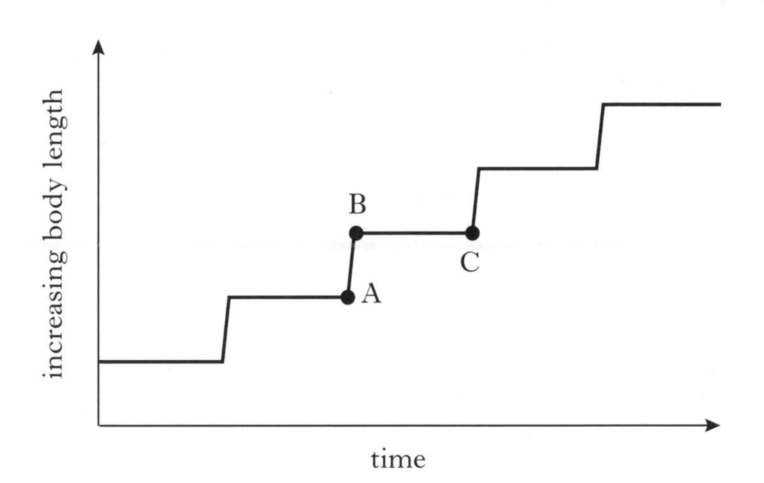


(*a*) Explain the growth pattern between points A and B and between points B and C shown on the graph.

A and B ____________________________________________

____________________________________________ **1**

B and C ____________________________________________

____________________________________________ **1**

(*b*) The diagram shows information about hormones involved in growth and development in humans.

pituitary gland

hormone X → increased amino acid uptake into bone

hormone Y → thyroxine released from gland Z

(i) Name hormones X and Y.

X ____________________________________________ **1**

Y ____________________________________________ **1**

(ii) Name gland Z.

____________________________________________ **1**

(iii) Describe the role of thyroxine in growth and development.

____________________________________________

____________________________________________ **1**

DO NOT WRITE IN THIS MARGIN

*Marks*

**12.** (*a*) In the control of lactose metabolism in *Escherichia coli (E. coli*), lactose acts as an inducer of the enzyme β-galactosidase.

The graph shows changes in concentrations of lactose and β-galactosidase after lactose was added to an *E. coli* culture growing in a container.

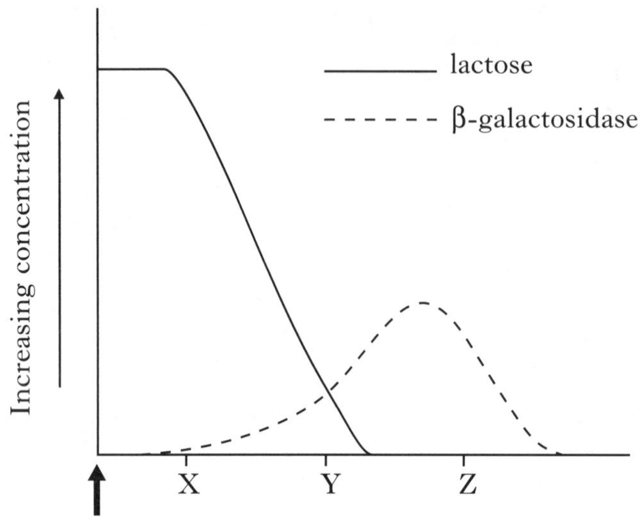


(i) Describe how the graph supports the statement that β-galactosidase breaks down lactose.

______________________________________________

______________________________________________ **1**

(ii) The statements in the table refer to times X, Y and Z on the graph.

Complete the table by writing **true** or **false** in each of the spaces provided.

| *Statement* | *True or False* |
|---|---|
| At time X the lactose is bound to the repressor | |
| At time Y the lactose is bound to the operator | |
| At time Z the repressor is bound to the operator | |

**2**

DO NOT WRITE IN THIS MARGIN

*Marks*

**12. (continued)**

(*b*) Complete the following sentences by underlining one of the alternatives in each pair.

Regeneration involves development of cells with specialised functions

from {differentiated / undifferentiated} cells through the switching on or off of

particular {hormones / genes}.

Mammals have {limited / extensive} powers of regeneration. **2**

**[Turn over**

DO NOT WRITE IN THIS MARGIN

*Marks*

**13.** The grid contains the names of substances that can influence growth and development in plants and animals.

| A | B | C | D |
|---|---|---|---|
| calcium | nitrogen | iron | phosphorus |
| E | F | G | H |
| vitamin D | magnesium | lead | potassium |

Use **letters from the grid** to answer the following questions.

Letters can be used once, more than once or not at all.

Each box should be completed using **one** letter only.

(*a*) Complete the table below.

| *Role in growth and development* | *Letter(s)* | |
|---|---|---|
| Important in membrane transport | | ■ |
| Present in chlorophyll | | |
| Present in nucleic acids | | |
| Needed for blood clotting | | ■ |

4

(*b*) Complete the sentence.

Deficiency of ☐ leads to rickets as a result of

poor ☐ absorption in the intestine.

1

*Marks*

**14.** The list below shows conditions which must be maintained within tolerable limits in the human body.

*List*

A blood glucose concentration
B blood water concentration
C body temperature

(*a*) Use **all** the letters from the list to complete the table below to show where each condition is monitored.

| *Hypothalamus* | *Pancreas* |
|---|---|
| | |

1

(*b*) The liver contains a reservoir of stored carbohydrate.

Name **two** hormones which can cause the breakdown of this carbohydrate to increase the concentration of glucose in the blood.

1 ____________________

2 ____________________ 1

(*c*) An increase in blood water concentration causes a reduction in the level of ADH in the bloodstream.

Describe the effect of this reduction on the kidney tubules.

____________________

____________________ 1

(*d*) (i) When body temperature falls below normal, the blood vessels in the skin respond.

State how the blood vessels in the skin respond and explain how this helps return body temperature to normal.

Blood vessel response ____________________ 1

Explanation ____________________

____________________ 1

(ii) What term is used to describe animals which derive most of their body heat from their own metabolism?

____________________ 1

DO NOT WRITE IN THIS MARGIN

*Marks*

## SECTION C

**Both questions in this section should be attempted.**

Note that each question contains a choice.

**Questions 1 and 2 should be attempted on the blank pages which follow.**

**Supplementary sheets, if required, may be obtained from the Invigilator.**

**All answers must be written clearly and legibly in ink.**

**Labelled diagrams may be used where appropriate.**

**1.** Answer **either** A **or** B.

**A.** Write notes on maintaining a water balance under the following headings:

(i) osmoregulation in **salt water** bony fish; **6**

(ii) water conservation in the desert rat. **4**

**(10)**

**OR**

**B.** Write notes on meiosis under the following headings:

(i) first and second meiotic divisions; **7**

(ii) its role in the production of new phenotypes. **3**

**(10)**

**In question 2, ONE mark is available for coherence and ONE mark is available for relevance.**

**2.** Answer **either** A **or** B.

**A.** Give an account of carbon fixation in photosynthesis and its importance to plants. **(10)**

**OR**

**B.** Give an account of the production of new viruses after the invasion of cells and the role of lymphocytes in cellular defence. **(10)**

[*END OF QUESTION PAPER*]

DO NOT WRITE IN THIS MARGIN

**SPACE FOR ANSWERS**

**SPACE FOR ANSWERS**

DO NOT WRITE IN THIS MARGIN

**SPACE FOR ANSWERS**

**SPACE FOR ANSWERS**

DO NOT WRITE IN THIS MARGIN

**SPACE FOR ANSWERS**

DO NOT WRITE IN THIS MARGIN

**SPACE FOR ANSWERS**

ADDITIONAL GRAPH PAPER FOR QUESTION 10(*f*)

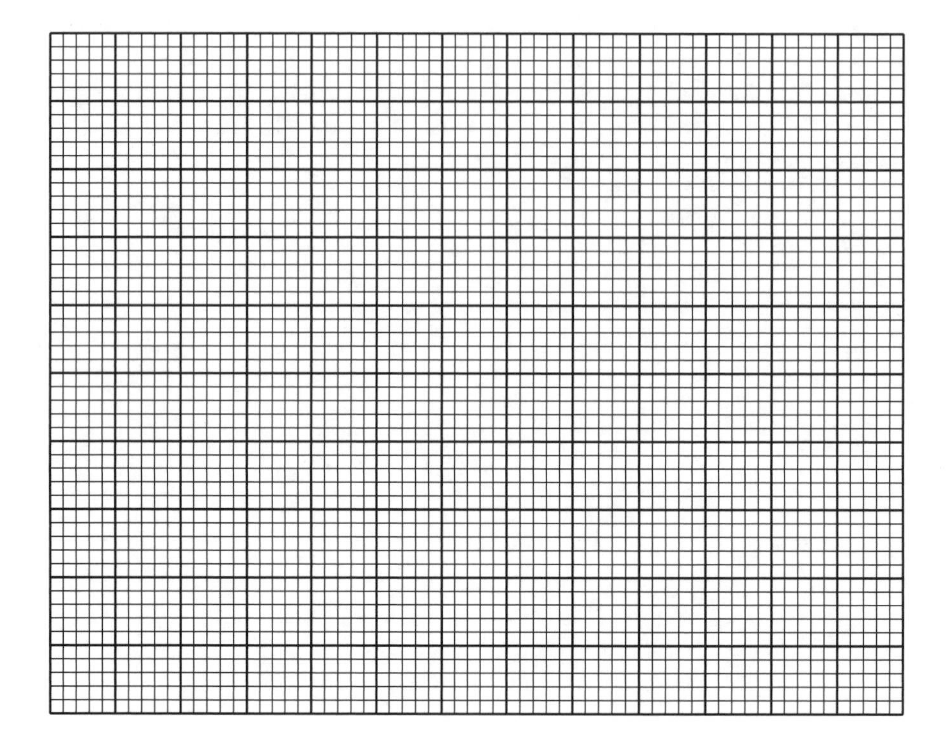

HIGHER

# 2012

HODDER GIBSON
LEARN MORE

**[BLANK PAGE]**

FOR OFFICIAL USE

| | | | | | |
|---|---|---|---|---|---|
| | | | | | |

Total for Sections B and C

# X007/12/02

NATIONAL QUALIFICATIONS 2012

WEDNESDAY, 23 MAY
1.00 PM – 3.30 PM

# BIOLOGY HIGHER

**Fill in these boxes and read what is printed below.**

Full name of centre

Town

Forename(s)

Surname

Date of birth
Day Month Year

Scottish candidate number

Number of seat

**SECTION A—Questions 1—30 (30 Marks)**

Instructions for completion of Section A are given on *Page two.*

For this section of the examination you must use an **HB pencil**.

**SECTIONS B AND C (100 Marks)**

1 (a) All questions should be attempted.

(b) It should be noted that in **Section C** questions 1 and 2 each contain a choice.

2 The questions may be answered in any order but all answers are to be written in the spaces provided in this answer book, **and must be written clearly and legibly in ink**.

3 Additional space for answers will be found at the end of the book. If further space is required, supplementary sheets may be obtained from the Invigilator and should be inserted inside the **front** cover of this book.

4 The numbers of questions must be clearly inserted with any answers written in the additional space.

5 Rough work, if any should be necessary, should be written in this book and then scored through when the fair copy has been written. If further space is required, a supplementary sheet for rough work may be obtained from the Invigilator.

6 Before leaving the examination room you must give this book to the Invigilator. If you do not, you may lose all the marks for this paper.

**Read carefully**

1 Check that the answer sheet provided is for **Biology Higher (Section A)**.

2 For this section of the examination you must use an **HB pencil**, and where necessary, an eraser.

3 Check that the answer sheet you have been given has **your name**, **date of birth**, **SCN** (Scottish Candidate Number) and **Centre Name** printed on it.

Do not change any of these details.

4 If any of this information is wrong, tell the Invigilator immediately.

5 If this information is correct, **print** your name and seat number in the boxes provided.

6 The answer to each question is **either** A, B, C or D. Decide what your answer is, then, using your pencil, put a horizontal line in the space provided (see sample question below).

7 There is **only one correct** answer to each question.

8 Any rough working should be done on the question paper or the rough working sheet, **not** on your answer sheet.

9 At the end of the examination, put the **answer sheet for Section A inside the front cover of this answer book**.

**Sample Question**

The apparatus used to determine the energy stored in a foodstuff is a

A calorimeter

B respirometer

C klinostat

D gas burette.

The correct answer is **A**—calorimeter. The answer **A** has been clearly marked in **pencil** with a horizontal line (see below).

A B C D

**Changing an answer**

If you decide to change your answer, carefully erase your first answer and using your pencil fill in the answer you want. The answer below has been changed to **D**.

A B C D

## SECTION A

**All questions in this section should be attempted.**

**Answers should be given on the separate answer sheet provided.**

**1.** The diagram below shows the arrangement of molecules in part of a cell membrane.

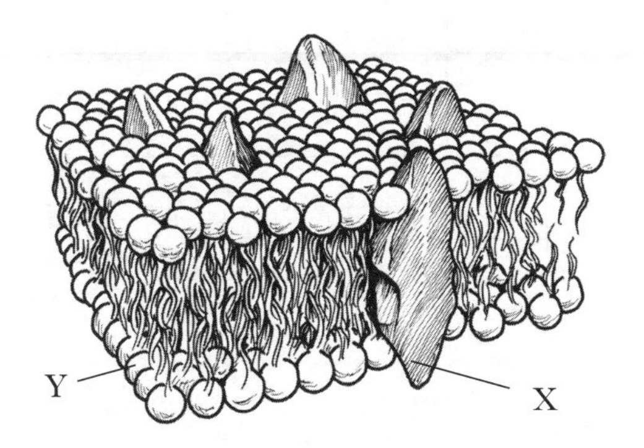


What types of molecule are represented by X and Y?

| | *X* | *Y* |
|---|---|---|
| A | Phospholipid | Protein |
| B | Protein | Phospholipid |
| C | Protein | Carbohydrate |
| D | Carbohydrate | Protein |

**2.** The experiment below was set up to demonstrate osmosis.

Visking tubing is selectively permeable.

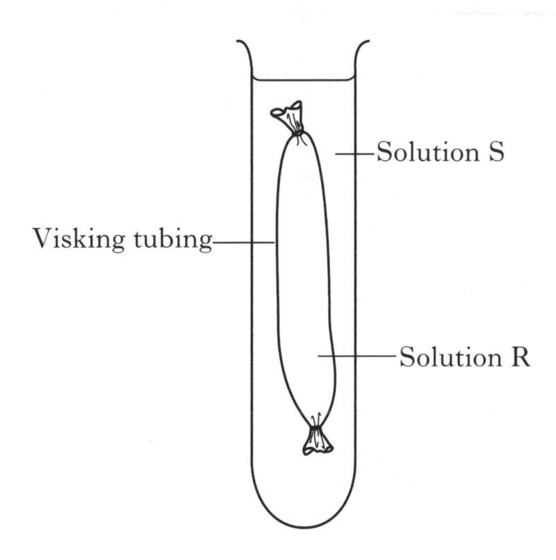


The following results were obtained.

Initial mass of Visking tubing + contents = 10·0 g

Mass of Visking tubing + contents after experiment = 8·2 g

The results shown above could be obtained when

A R is a 5% salt solution and S is a 10% salt solution

B R is a 10% salt solution and S is a 5% salt solution

C R is a 10% salt solution and S is water

D R is a 5% salt solution and S is water.

[Turn over

3. The diagram below refers to the plasma membrane of an animal cell.

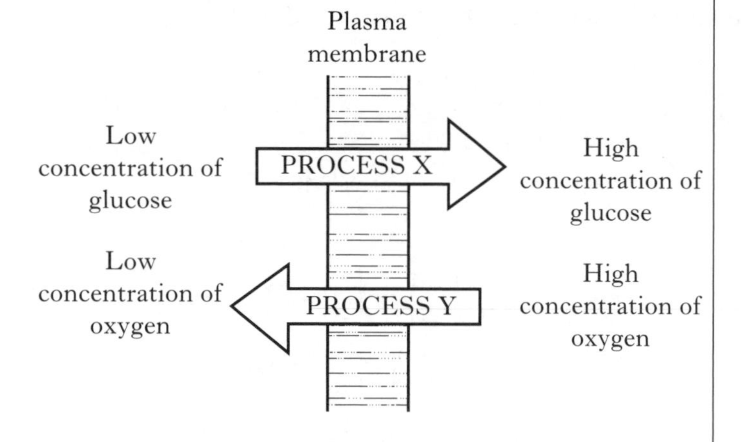


Identify the two processes X and Y.

| | *X* | *Y* |
|---|---|---|
| A | active transport | diffusion |
| B | diffusion | active transport |
| C | respiration | diffusion |
| D | active transport | respiration |

4. The following absorption spectra were obtained from four different plant extracts. Black areas indicate light which has been absorbed by the extracts.

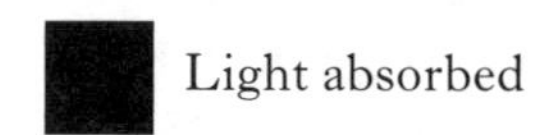


A 

B 

C 

D 

Which extract contains chlorophyll?

5. The graph below shows changes in the mass of chlorophyll and rate of photosynthesis in leaves during a 10 day period in autumn.

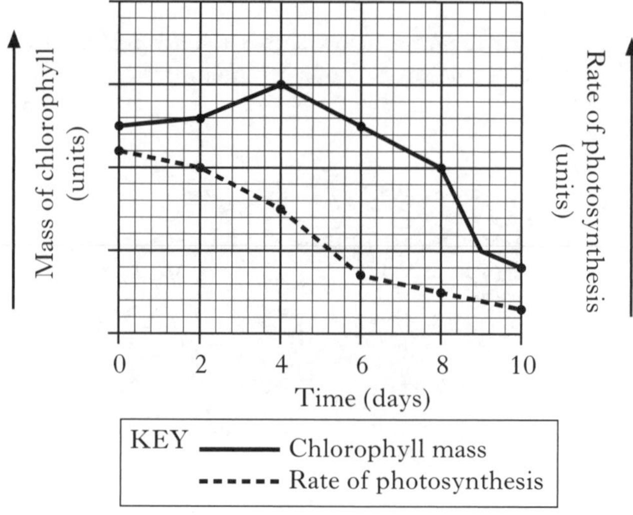


Chlorophyll content of leaves can limit the rate of photosynthesis.

During which period do the results **not** support this statement?

A 0–4 days

B 4–8 days

C 8–9 days

D 9–10 days

6. The processes in the list below occur in living cells.

1 NADP acts as a hydrogen acceptor.

2 ATP is synthesised.

3 Oxygen acts as a hydrogen acceptor.

4 Carbon dioxide enters a cycle of reactions.

Which line of the table below matches each process with the set of reactions in which it occurs?

| | *Set of reactions* | | |
|---|---|---|---|
| | *Respiration only* | *Photosynthesis only* | *Respiration and photosynthesis* |
| A | 2 | 1 and 4 | 3 |
| B | 3 | 1 and 4 | 2 |
| C | 2 | 4 | 1 and 3 |
| D | 3 | 4 | 1 and 2 |

**7.** The statements in the list below refer to respiration.

1 Carbon dioxide is released.

2 Occurs during aerobic respiration.

3 The end product is pyruvic acid.

4 The end product is lactic acid.

Which statements describe glycolysis?

A 1 and 4

B 1 and 3

C 2 and 3

D 2 and 4

**8.** The graph below shows changes in food stores in a human body during four weeks without food.

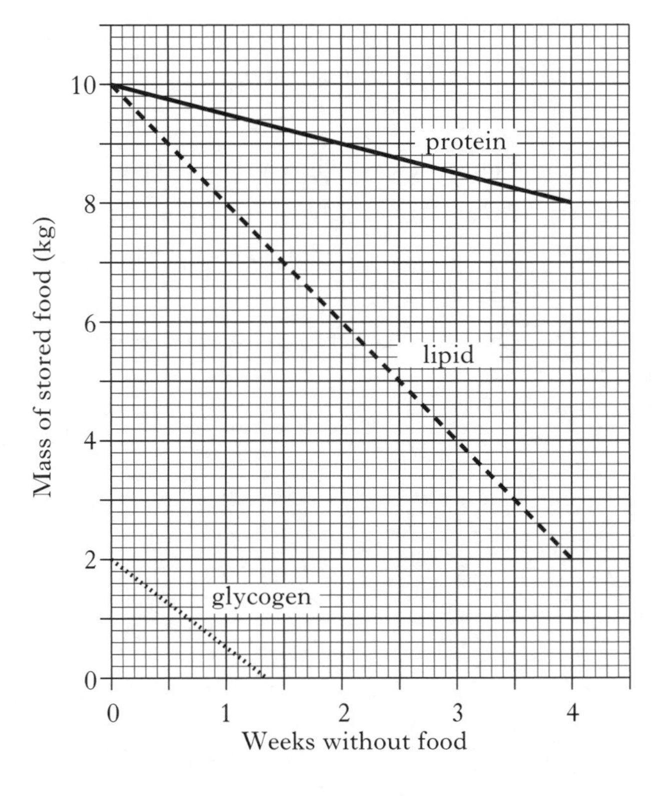


Which of the following conclusions can be drawn from the graph?

A Each food store decreases at the same rate during week one.

B Between weeks three and four the body gains most energy from protein.

C The lipid food store decreases at a faster rate than the other food stores during week one.

D Between weeks one and four, the body only gains energy from lipid and protein.

**9.** A fragment of DNA was found to consist of 72 nucleotide base pairs. What is the total number of deoxyribose sugars in this fragment?

A 24

B 36

C 72

D 144

**10.** Insulin synthesised in a pancreatic cell is secreted. Its route from synthesis to secretion includes

A Golgi apparatus → endoplasmic reticulum → ribosome

B ribosome → Golgi apparatus → endoplasmic reticulum

C endoplasmic reticulum → ribosome → Golgi apparatus

D ribosome→ endoplasmic reticulum → Golgi apparatus.

**11.** The following diagram shows a pair of homologous chromosomes and the positions of 4 genes.

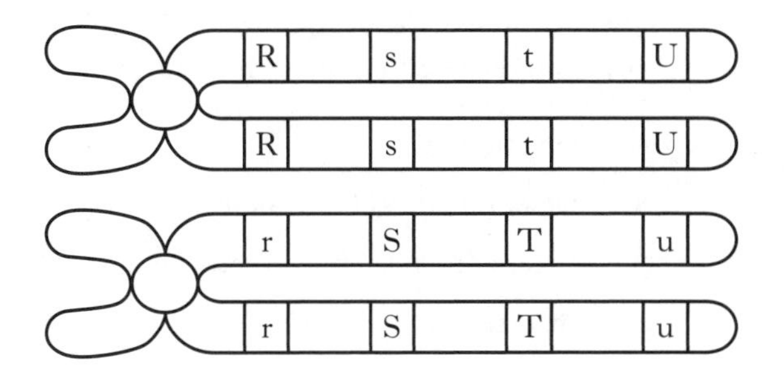


Between which of the following alleles would chiasma formation occur least often?

A r and t

B r and U

C r and s

D s and u

**[Turn over**

12. White eye colour in *Drosophila* is caused by a recessive sex-linked allele. The dominant allele is for red eyes.

What result would be obtained from a cross between a white-eyed female and a red-eyed male?

A All white-eyed flies

B All red-eyed flies

C Equal numbers of white-eyed and red-eyed flies.

D Three times as many red-eyed flies as white-eyed flies

13. Which of the following is true of polyploid plants?

A They have reduced yield and the diploid chromosome number.

B They have increased yield and the diploid chromosome number.

C They have reduced yield and sets of chromosomes greater than diploid.

D They have increased yield and sets of chromosomes greater than diploid.

14. Which of the following gene mutations alters all the amino acids in the protein being coded for, from the position of the mutation?

A Deletion and insertion

B Deletion and substitution

C Inversion and substitution

D Insertion and inversion

15. The following steps are involved in the process of genetic engineering.

1 Insertion of a plasmid into a bacterial host cell.

2 Use of an enzyme to cut out the desired gene from a chromosome.

3 Insertion of the desired gene into the bacterial plasmid.

4 Use of an enzyme to open a bacterial plasmid.

What is the correct sequence of these steps?

A 4 1 2 3

B 2 4 3 1

C 4 3 1 2

D 2 3 4 1

16. Which of the following is the function of cellulase in the process of somatic fusion in plants?

A Conversion of cells to protoplasts

B Isolation of cells from the parent plants

C Fusion of protoplasts from different plants

D Callus formation from hybrid protoplasts

**17.** The drinking rate and concentrations of sodium and chloride ions in blood were measured over a six hour period after a salmon was transferred from freshwater to sea water. The results are shown in the graph below.

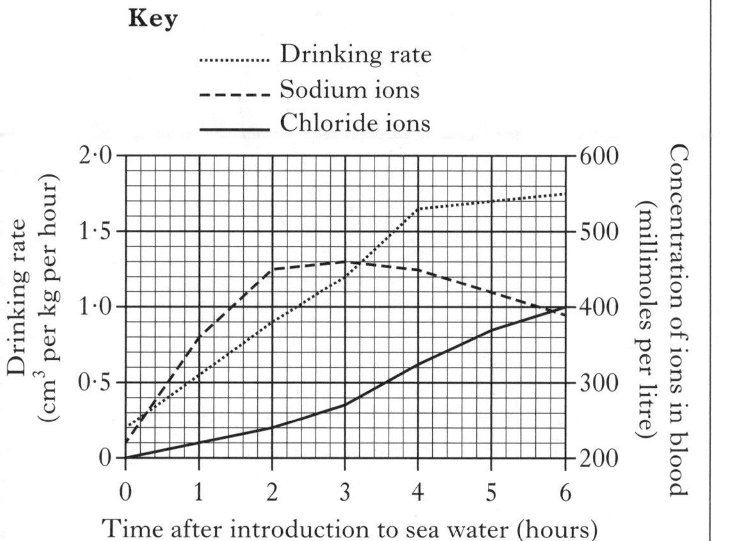


Which line in the table correctly shows the results after three hours?

| | *Drinking rate* ($cm^3$ per kg per hour) | *Sodium ion concentration* (millimoles per litre) | *Chloride ion concentration* (millimoles per litre) |
|---|---|---|---|
| A | 1·4 | 270 | 460 |
| B | 1·2 | 460 | 270 |
| C | 1·2 | 270 | 460 |
| D | 1·4 | 460 | 270 |

**18.** Under which of the following conditions is the rate of transpiration in a plant likely to be the highest?

| | *Wind speed* | *Air temperature* | *Air humidity* |
|---|---|---|---|
| A | high | high | high |
| B | high | high | low |
| C | high | low | low |
| D | low | high | high |

**19.** Which line in the table below shows features likely to be found in a plant and in a small mammal **both** adapted to life in hot desert conditions?

| | *Plant* | *Small mammal* |
|---|---|---|
| A | reduced root system | large number of sweat glands |
| B | rolled leaves | large number of glomeruli |
| C | small number of stomata | nocturnal habit |
| D | succulent tissues | short kidney tubules |

**20.** An investigation was set up to demonstrate the response of flatworms to the presence of food. A piece of liver and a glass bead were placed in a dish. 15 flatworms were then scattered randomly into the dish as shown in the diagram below.

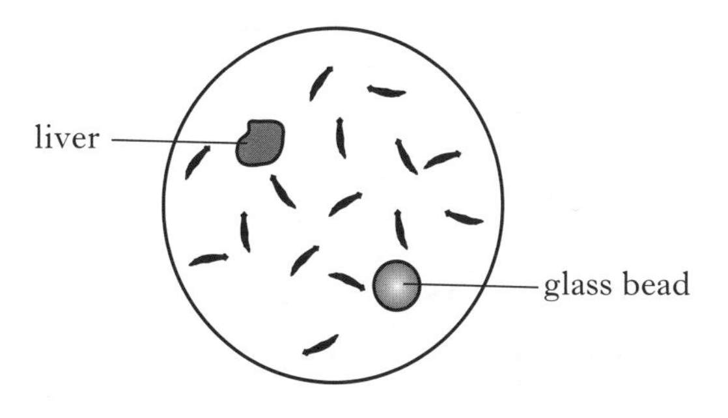


The purpose of the glass bead is to show that flatworms

A use cooperative foraging behaviour

B respond differently to food compared to other objects

C move randomly in search of food

D move towards any large object in the search for food.

[Turn over

**21.** The graph below shows the growth, in length, of a human fetus during pregnancy.

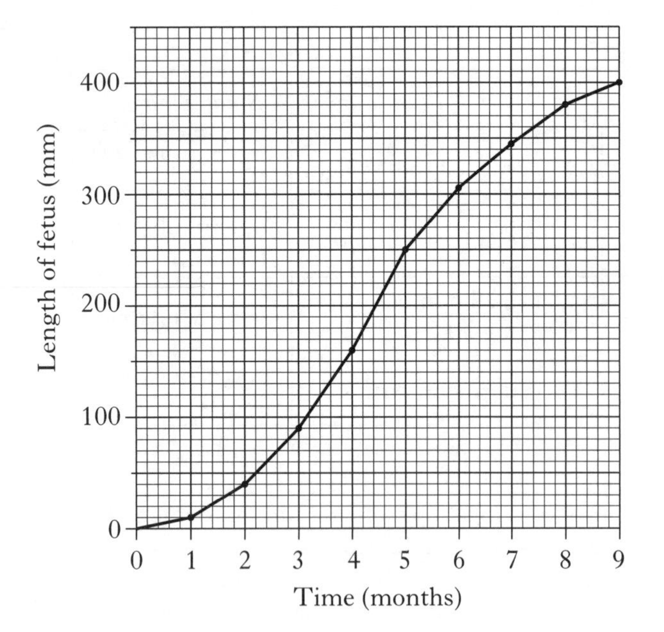


What is the percentage increase in length of the fetus during the final 4 months of pregnancy?

A 33·3

B 60·0

C 62·5

D 150·0

**22.** The diagram below shows a section of a woody twig.

Identify the position of a meristem.

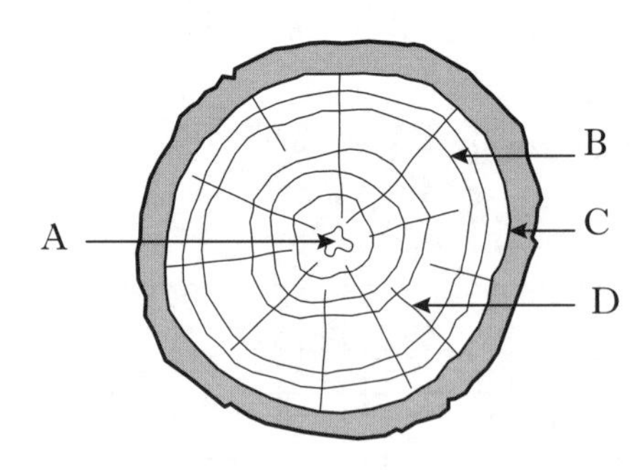


**23.** In which of the following processes does gibberellic acid (GA) have a role during the growth and development of plants?

A Breaking dormancy

B Root formation in cuttings

C Leaf abscission

D Apical dominance

**24.** The table below contains descriptions of terms used to illustrate the control of lactose metabolism in the bacterium *Esherichia coli*.

Which line in the table contains terms which correctly match the descriptions given?

| | *Description* | | | |
|---|---|---|---|---|
| | *Produces lactose digesting enzyme* | *Acts as the inducer* | *Produces repressor molecule* | *Switches on structural gene* |
| A | regulator gene | repressor molecule | structural gene | lactose |
| B | structural gene | repressor molecule | regulator gene | lactose |
| C | regulator gene | lactose | structural gene | operator |
| D | structural gene | lactose | regulator gene | operator |

**25.** An investigation into the influence of different concentrations of IAA on the development of certain plant organs was carried out. The growth-inhibiting or growth-promoting effects are shown below.

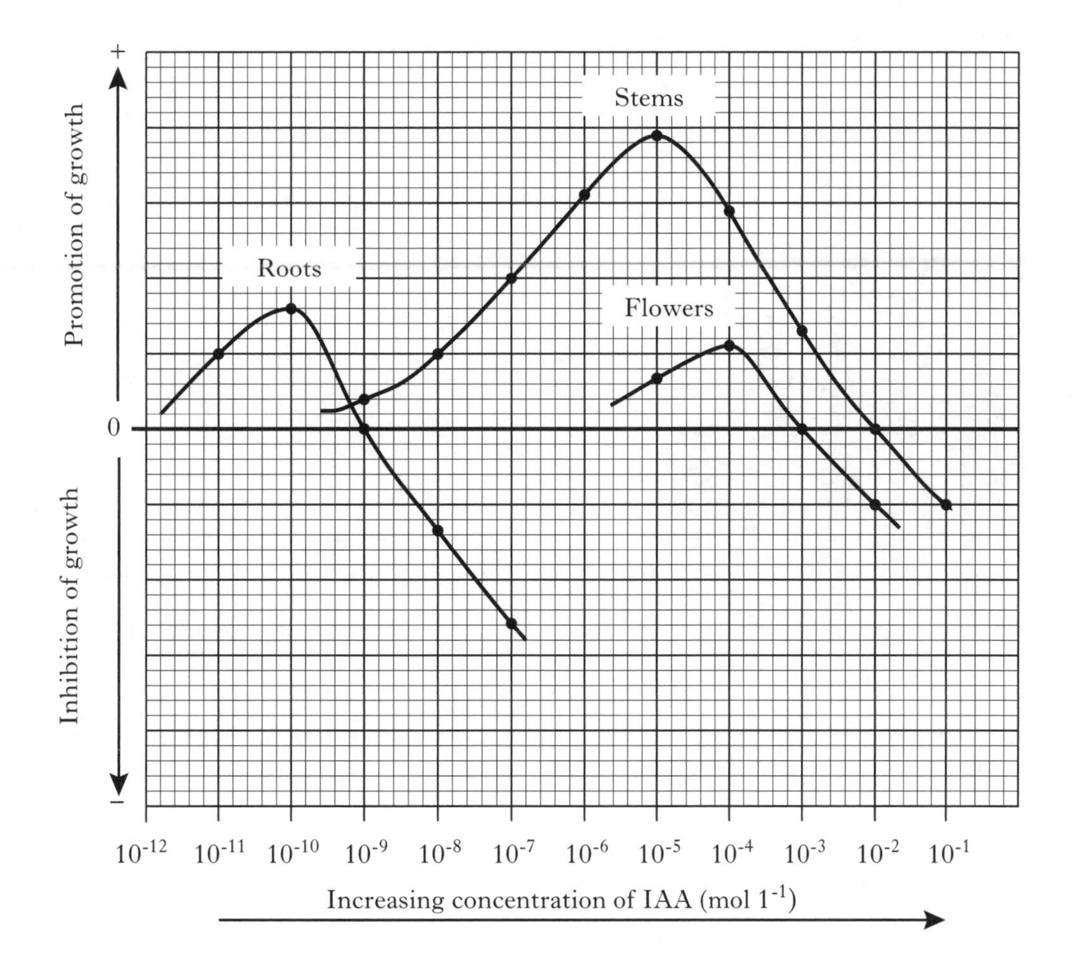


The graph shows that an IAA concentration of

A $10^{-3}$ mol $l^{-1}$ promotes flower and stem growth

B $10^{-5}$ mol $l^{-1}$ promotes stem and flower growth

C $10^{-7}$ mol $l^{-1}$ promotes root and stem growth

D $10^{-9}$ mol $l^{-1}$ inhibits stem growth and promotes root growth.

**[Turn over**

**26.** The bar chart shows the units of vitamin D provided by 100 g of various foods and the graph shows the number of units required daily by humans of different ages.

**Bar chart**

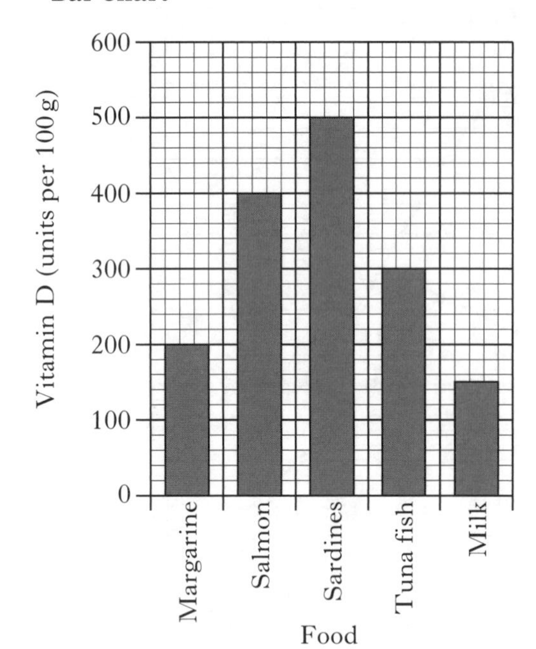


**Graph**

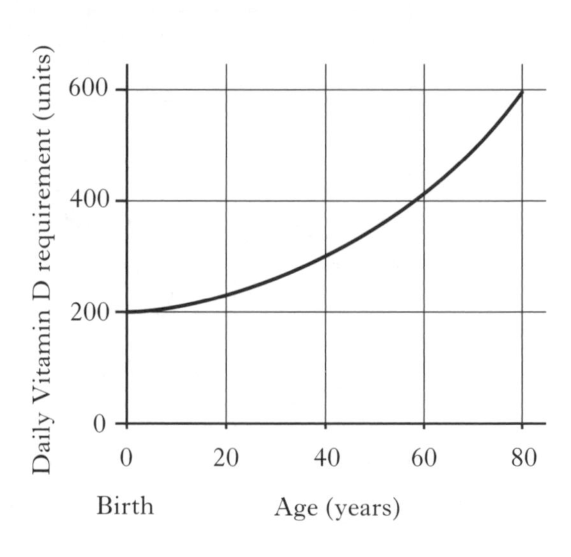


At what age would eating 100 g of tuna fish and 100 g of margarine exactly provide the number of units of vitamin D required in one day?

A 0 (birth)

B 40

C 65

D 70

---

**27.** The list below shows effects of various drugs on fetal development.

1 Reduced growth

2 Limb deformation

3 Reduced mental development.

Which effects are associated with the intake of nicotine during pregnancy in humans?

A 1 only

B 1 and 2 only

C 1 and 3 only

D 1, 2 and 3

**28.** Which of the following best defines etiolation?

A The inhibition of development of lateral buds

B The result of a magnesium deficiency in seedlings

C The growth of a stem towards directional light

D The effect on seedlings of being grown in the dark

**29.** An effect of a high concentration of antidiuretic hormone (ADH) on the kidney is to

A increase tubule permeability which increases water reabsorption

B decrease tubule permeability which prevents excessive water loss

C increase glomerular filtration rate which increases urine production

D decrease glomerular filtration rate which reduces urine production.

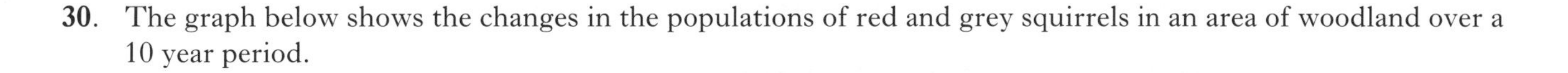


**30.** The graph below shows the changes in the populations of red and grey squirrels in an area of woodland over a 10 year period.

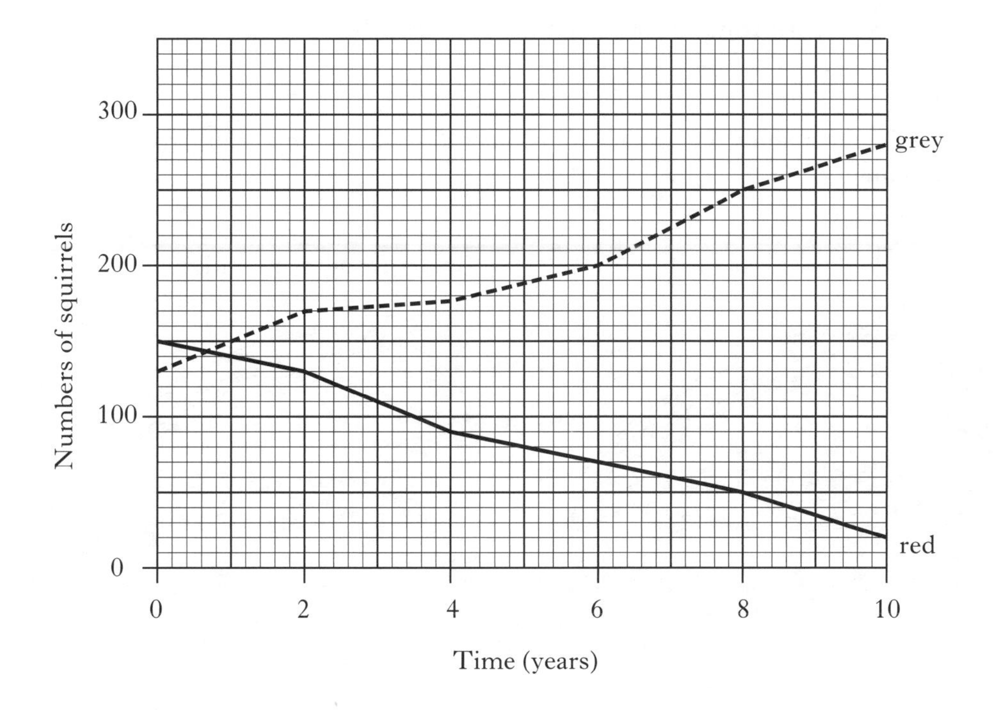


From the graph the following conclusions were suggested.

1 The grey squirrel population increases by 150% over the ten year period.

2 The red squirrel numbers decreased from 150 to 20 over the ten year period.

3 After eight years the grey squirrel population was five times greater than the red.

Which of the conclusions are correct?

A 1 and 2 only

B 1 and 3 only

C 2 and 3 only

D 1, 2 and 3

**Candidates are reminded that the answer sheet MUST be returned INSIDE the front cover of this answer book.**

**[Turn over**

DO NOT WRITE IN THIS MARGIN

## SECTION B

*Marks*

**All questions in this section should be attempted.**

**All answers must be written clearly and legibly in ink.**

**1.** (*a*) The diagrams show a normal chloroplast and one from a plant treated with a weedkiller.

Normal chloroplast

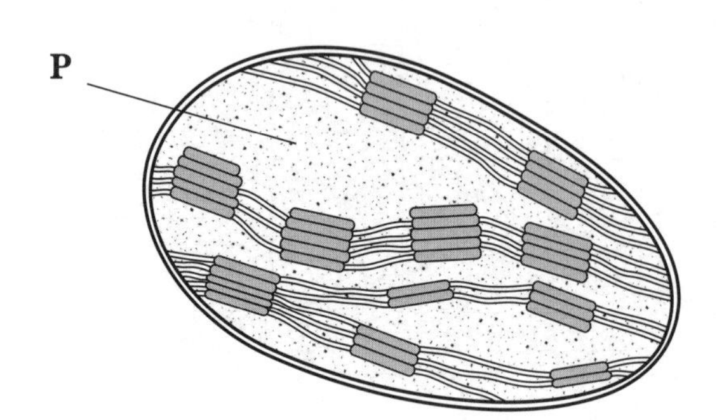


Chloroplast from treated plant

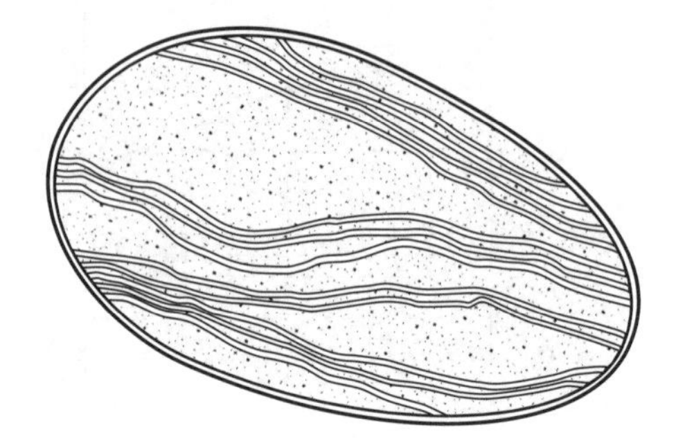

(i) Name area **P**.

______________________________ **1**

(ii) Describe how the structure of the chloroplast from the treated plant has been affected by the weedkiller.

______________________________

______________________________ **1**

(iii) The production of two substances required for the carbon fixation stage (Calvin cycle) was significantly decreased in the treated plant.

Name these **two** substances.

1 ______________________________

2 ______________________________ **2**

DO NOT WRITE IN THIS MARGIN

**1. (continued)**

*Marks*

(*b*) In an investigation into the carbon fixation stage of photosynthesis, algal cells were kept in a constant light intensity at 20 °C. The concentration of ribulose bisphosphate (RuBP) and glycerate phosphate (GP) in the cells was measured through the investigation.

The concentration of carbon dioxide available was changed from 1% to 0·003% as shown on the graph.

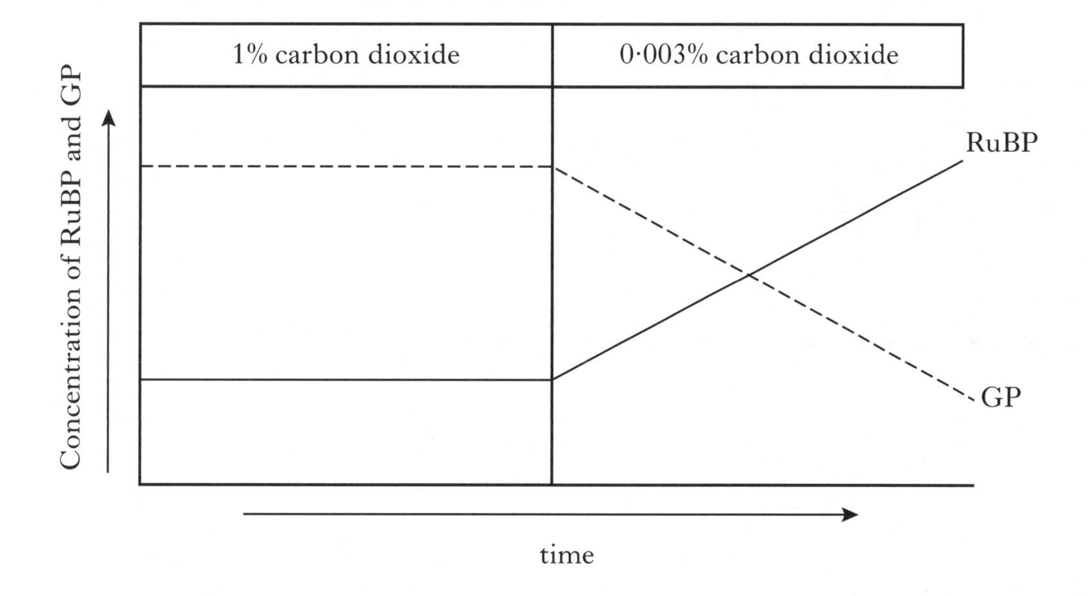


Explain the increase in RuBP concentration shown on the graph after the carbon dioxide concentration was reduced from 1% to 0·003%.

______________________________________________________________

______________________________________________________________

______________________________________________________________

______________________________________________________________ **2**

**[Turn over**

DO NOT WRITE IN THIS MARGIN

*Marks*

**2.** The diagram shows a stage in the synthesis of a protein.

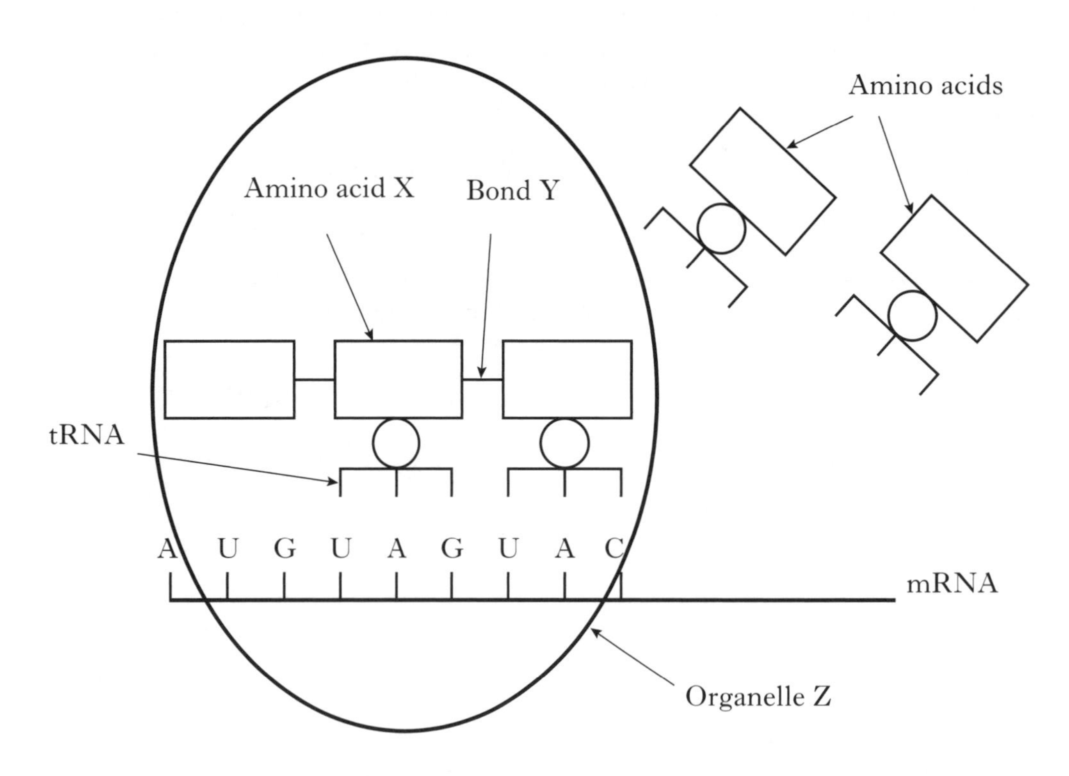


(*a*) Complete the diagram below by adding the appropriate letters to show the sequence of nine bases on the DNA strand from which the mRNA strand shown has been transcribed.

**1**

(*b*) Name organelle Z.

________________________________ **1**

(*c*) Give the anticodon which would be found on the tRNA carrying amino acid X in the diagram.

____________________ **1**

(*d*) Name bond Y.

____________________________ **1**

DO NOT WRITE IN THIS MARGIN

*Marks*

**3.** An investigation was carried out to study the effects of the concentration of sucrose solutions on pieces of tulip stem 45 mm in length. The pieces were placed in different concentrations of sucrose solution and measured after two hours of immersion.

The results are shown in the table below.

| *Sucrose concentration* (moles per litre) | *Length after 2 hours* (mm) |
|---|---|
| 0·2 | 50 |
| 0·3 | 48 |
| 0·4 | 46 |
| 0·5 | 44 |
| 0·6 | 42 |
| 0·7 | 42 |
| 0·8 | 42 |

(*a*) Explain the effect of the 0·2 moles per litre sucrose solution on the length of the pieces of the tulip stem.

______________________________________________

______________________________________________

______________________________________________ **1**

(*b*) Use information from the table to predict the concentration of a sucrose solution isotonic to the cells in the tulip stem.

______________________ moles per litre **1**

(*c*) Give the term which would be used to describe the cells in the tulip stem after immersion in a solution with a sucrose concentration of 0·7 moles per litre.

______________________________ **1**

**[Turn over**

DO NOT WRITE IN THIS MARGIN

*Marks*

4. The diagram shows apparatus used in an investigation of aerobic respiration in snails.

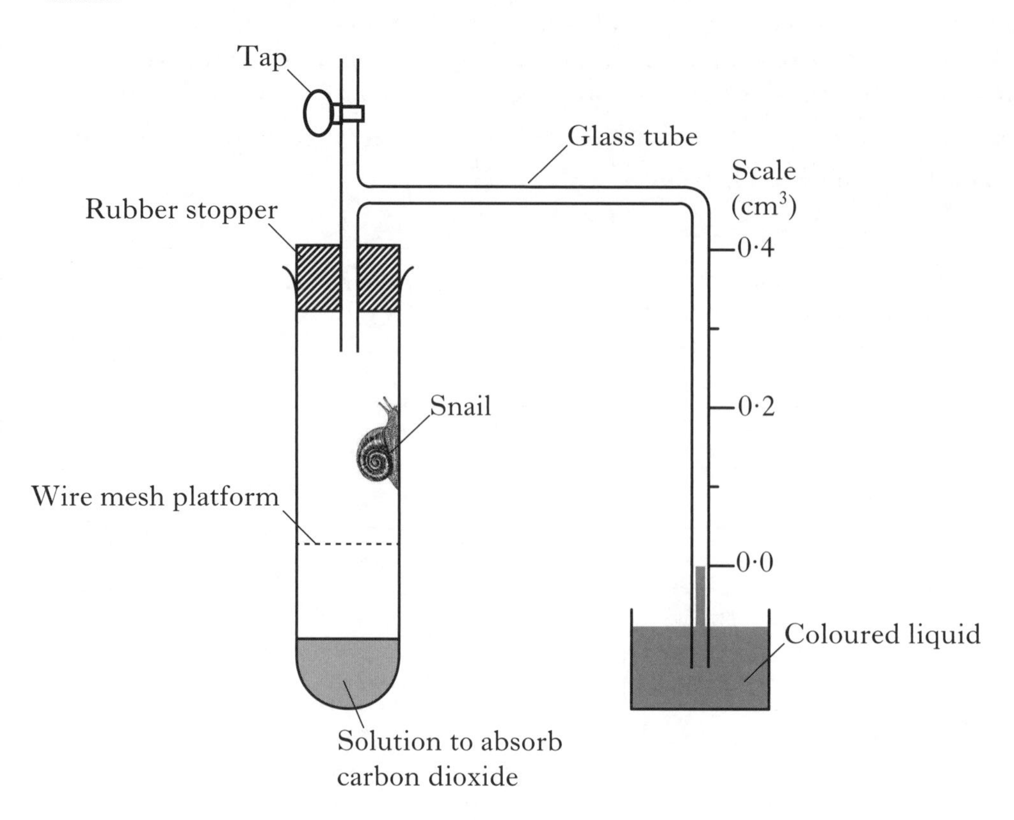


The tap was kept open to the air for 15 minutes, and to start the experiment the tap was closed and the reading on the scale recorded. Every 2 minutes for 10 minutes the reading on the scale was again recorded and the results shown in the table below. The apparatus was kept at 20 °C throughout.

| *Time after tap closed* (minutes) | *Reading on scale* ($cm^3$) |
|---|---|
| 0 | 0·00 |
| 2 | 0·04 |
| 4 | 0·08 |
| 6 | 0·12 |
| 8 | 0·16 |
| 10 | 0·20 |

(*a*) State why the apparatus was left for 15 minutes with the tap open before readings were taken.

______________________________ **1**

(*b*) Describe a suitable control for this investigation.

______________________________

______________________________ **1**

DO NOT WRITE IN THIS MARGIN

**4. (continued)**

*Marks*

(*c*) To increase the reliability of results, the experiment was repeated several times. Identify **one** variable, not already mentioned, that would have to be kept the same each time to ensure that the procedure was valid.

________________________________________ **1**

(*d*) On the grid below, draw a line graph to show the reading on the scale against time, choosing appropriate scales so that the graph fills most of the grid.

(Additional graph paper, if required, will be found on *Page forty*.)

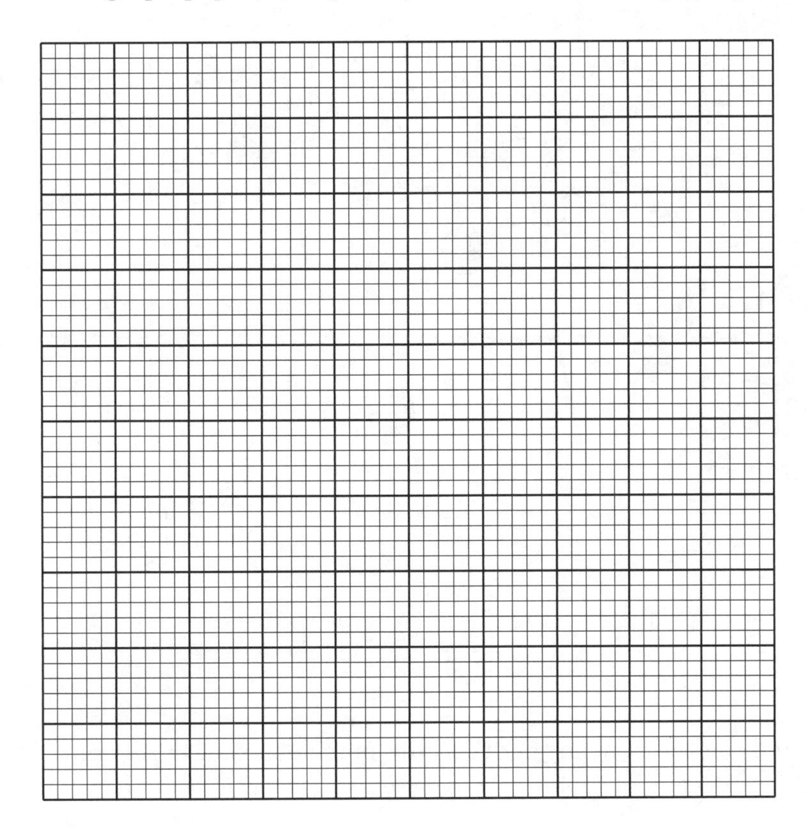

**2**

(*e*) The mass of the snail was 5·0 g.

Use results in the table to calculate the rate of oxygen uptake by the snail over the 10 minute period.

*Space for calculation*

____________________ $cm^3$ oxygen per minute per gram of snail **1**

(*f*) Explain how the respiration of the snail and the presence of the solution in the apparatus accounts for the movement of the coloured liquid on the scale.

________________________________________

________________________________________

________________________________________

________________________________________ **2**

DO NOT WRITE IN THIS MARGIN

*Marks*

**5.** (*a*) Some clover plants are cyanogenic. These plants discourage grazing by releasing cyanide when their leaves are damaged by invertebrate herbivores. The map shows four zones of Europe with their average January temperatures. The pie-charts represent the percentages of cyanogenic and non-cyanogenic clover plants at sample sites in these zones.

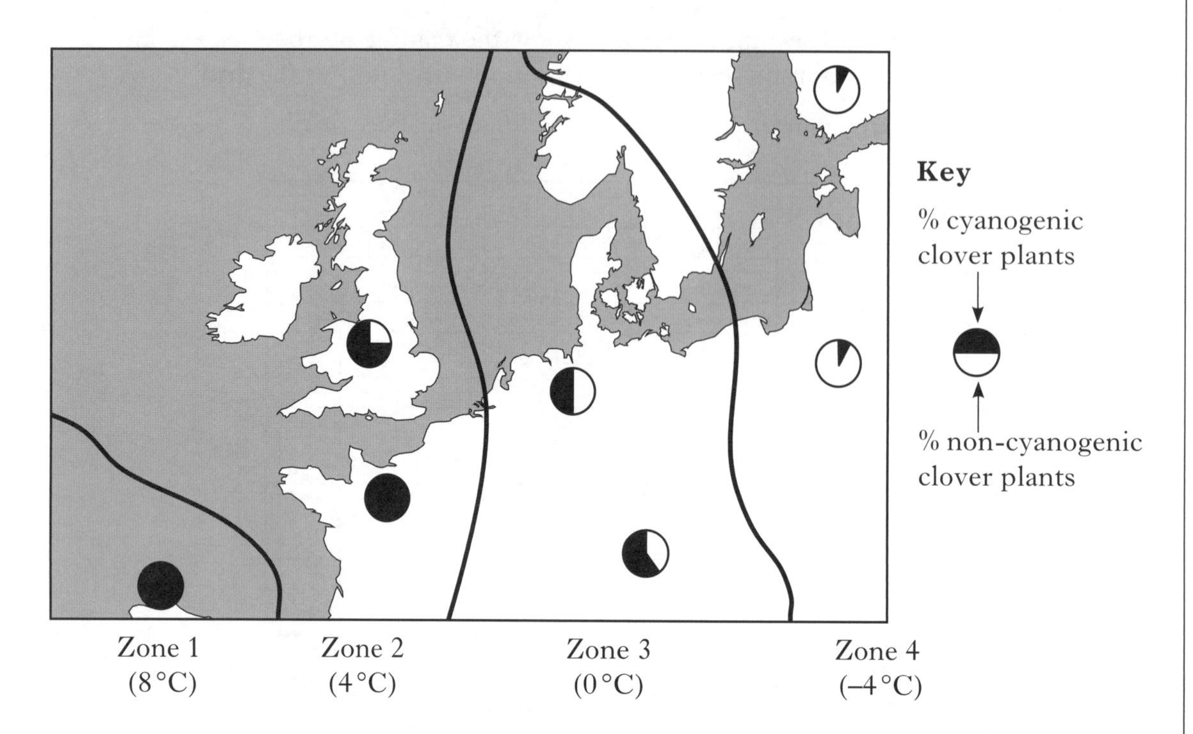


Zones with their average January temperatures

(i) Describe the relationship between the percentages of cyanogenic clover plants and the average January temperature of the zone in which they occur.

______________________________________________

______________________________________________ **1**

(ii) Zones with higher average January temperatures have higher densities of invertebrate herbivores.

Explain how this accounts for the distribution of the different clover varieties.

______________________________________________

______________________________________________

______________________________________________ **1**

DO NOT WRITE IN THIS MARGIN

Marks

**5. (a) (continued)**

(iii) Apart from cyanide, name **one** other toxic compound produced by plants to discourage grazing by herbivores.

________________________________________ **1**

(iv) State **one** feature of some plant species which allows them to tolerate grazing by herbivores.

__________________________________________________________ **1**

(b) Name a substance produced by some plants which acts as a barrier to prevent the spread of infection from a wound site.

________________________________________ **1**

**[Turn over**

DO NOT WRITE IN THIS MARGIN

*Marks*

**6.** (*a*) Peregrine falcons are predators which hunt wading birds such as redshank. In an investigation, the hunting success of peregrines and the sizes and distances of the feeding flocks of redshank from cover were recorded. The results are shown on the graph below.

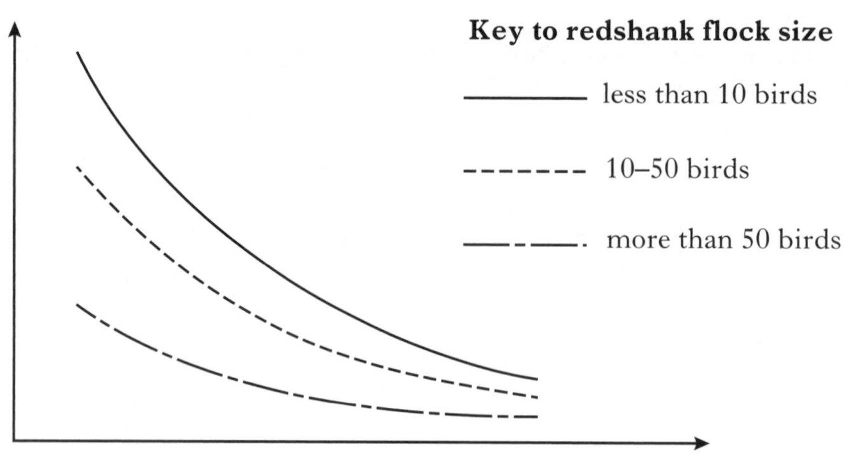


Give **two** conclusions which can be drawn from the results.

1 ____________________________________________

____________________________________________ **1**

2 ____________________________________________

____________________________________________ **1**

(*b*) Hermit crabs withdraw into their shells if disturbed by small stones dropped into the water. If this harmless stimulus is repeated, the number of crabs responding decreases.

Name the type of behaviour shown by the crabs when they:

(i) withdraw into their shells;

____________________________________________ **1**

(ii) no longer respond to the repeated harmless stimulus.

____________________________________________ **1**

DO NOT WRITE IN THIS MARGIN

*Marks*

**7.** (*a*) The diagram shows a stage in meiosis in a cell from a Hawkweed plant.

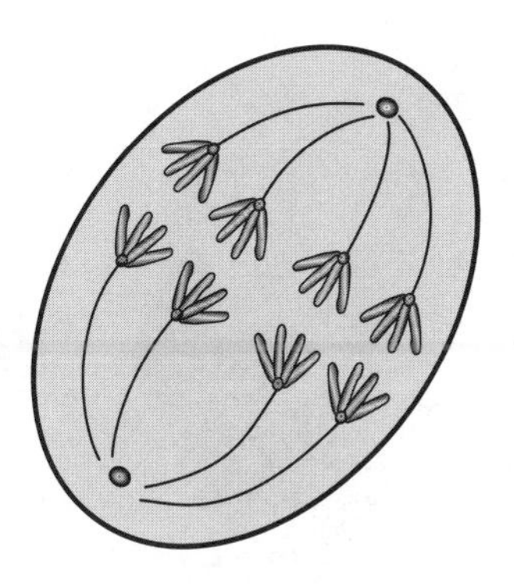

(i) Name a structure in a Hawkweed flower in which meiosis occurs.

_______________________ **1**

(ii) In the following sentence, underline the word in the choice brackets which identifies the stage of meiosis shown and give a reason for your choice.

This cell is from the { first / second } meiotic divison.

Reason _______________________

_______________________ **1**

(iii) State the number of chromosomes which would be found in a gamete and in a gamete mother cell from this plant.

gamete ____________

gamete mother cell ____________ **1**

(*b*) Mutation during meiosis can lead to new phenotypes.

(i) Other than mutation, state **one** feature of meiosis which can lead to the production of new phenotypes.

_______________________ **1**

(ii) Name the process which could result in the presence of an **extra** chromosome in a gamete.

_______________________ **1**

[Turn over

DO NOT WRITE IN THIS MARGIN

*Marks*

**8.** Comb shape in chickens is determined by two genes located on **different** chromosomes. One of these genes has alleles **A** and **a** and the other has alleles **B** and **b**.

Single comb

Rose comb

Pea comb

Cushion comb

- Chickens without alleles **A** or **B** have single combs
- Chickens with allele **B** but not **A** have rose combs
- Chickens with allele **A** but not **B** have pea combs
- Chickens with alleles **A** and **B** have cushion combs

(*a*) A male heterozygous for both genes was crossed with a female with a single comb.

(i) Complete the table below to show the parent genotypes and phenotypes and the genotypes of their gametes.

| | *Male* | *Female* |
|---|---|---|
| Parent genotypes | AaBb | |
| Parent phenotypes | | Single comb |
| Genotype(s) of gametes | | ab |

2

(ii) Give the expected ratio of phenotypes for the offspring in this cross.

*Space for working*

________ : ________ : ________ : ________

cushion comb rose comb pea comb single comb

1

(*b*) State the term used to describe genes which are found on the **same** chromosome.

________________________________________

1

DO NOT WRITE IN THIS MARGIN

*Marks*

**9.** In an investigation, 50 salmon were kept in a tank of fresh water for four days, then transferred to a tank of salt water for a further six days.

Each day, their gills were examined and the average diameter of chloride secretory cells was recorded.

The results are shown in the table below.

| *Contents of tank* | *Day* | *Average diameter of chloride secretory cells* (micrometres) |
|---|---|---|
| Fresh water | 1 | 203 |
| | 2 | 204 |
| | 3 | 202 |
| | 4 | 203 |
| Salt water | 5 | 280 |
| | 6 | 365 |
| | 7 | 471 |
| | 8 | 557 |
| | 9 | 615 |
| | 10 | 615 |

(*a*) In the following sentence, underline one alternative in each pair to make the sentence correct.

On day 3 the salmon are {hypertonic / hypotonic} to their surroundings and their chloride secretory cells actively transport salts {into / out of} the salmon. **1**

(*b*) Describe the effect of salt water on the average diameter of the chloride secretory cells between day 5 and day 10.

________________________________

________________________________ **2**

(*c*) Following the investigation, the fish were returned to fresh water. In the table below, tick (✓) **one** box in each row to show how this change affects kidney function.

| *Kidney Function* | *Increases* | *Decreases* | *Stays the same* |
|---|---|---|---|
| Filtration rate | | | |
| Urine concentration | | | |
| Urine volume | | | |

**2**

**10.** French bean plants were grown over a period of four weeks in solutions containing different concentrations of lead ions.

After this period, measurements of transpiration rate, dry mass and lead content were taken from plants grown in each solution.

The **Graph** shows the average transpiration rate.

The **Table** shows the average dry masses of the roots and shoots.

The **Bar Chart** shows the average lead content of the roots and shoots.

**Graph**

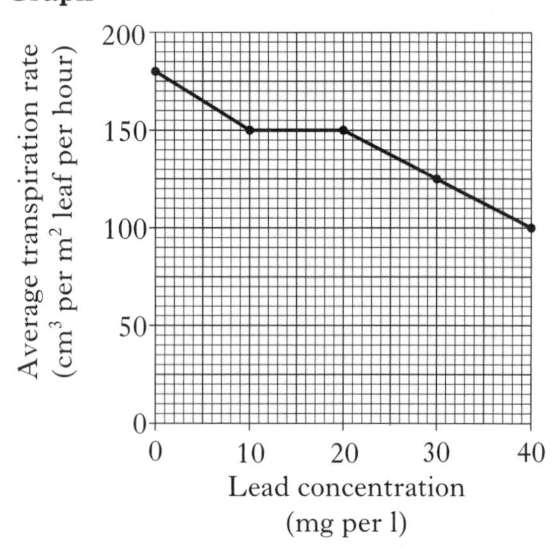


**Table**

| *Lead concentration* (mg per l) | *Average dry mass of roots* (g) | *Average dry mass of shoots* (g) |
|---|---|---|
| 0 | 3·1 | 0·4 |
| 10 | 3·2 | 0·3 |
| 20 | 2·2 | 0·2 |
| 30 | 2·0 | 0·1 |
| 40 | 1·3 | 0·1 |

**Bar Chart**

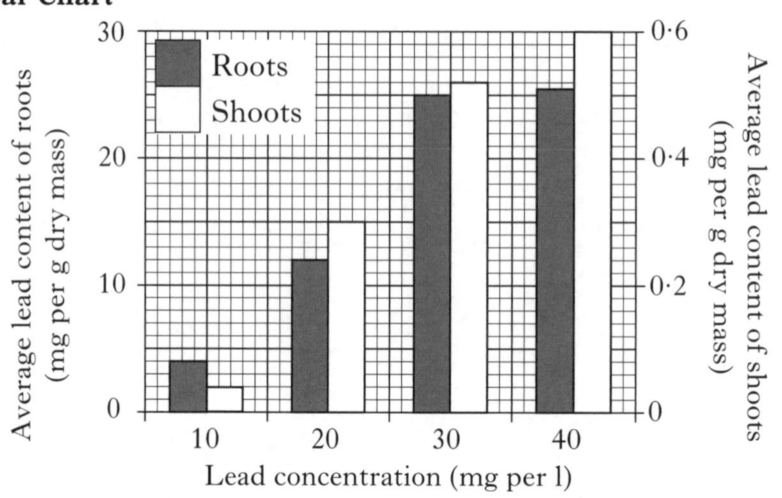


DO NOT WRITE IN THIS MARGIN

*Marks*

(*a*) (i) **Use values from the Graph** to describe the changes in average transpiration rate as the lead concentration increases from 0 to 40 mg per litre. **2**

(ii) What evidence from the **Graph** suggests that lead concentration is **not** the only factor affecting transpiration rate in this investigation? **1**

**10. (continued)** *Marks*

(*b*) Use information from the **Table**, to calculate the percentage decrease in **combined** average dry mass of shoots **and** roots when the lead concentration in the solution was increased from 0 to 40 mg per litre.

*Space for calculation*

________________ % decrease **1**

(*c*) Use information from the **Bar Chart** to:

(i) give the lead content of the shoots when the lead concentration of the solution was 10 mg per litre;

________________ mg per g dry mass **1**

(ii) calculate the simplest whole number ratio of lead content in roots to shoots for plants grown in a solution of 20 mg lead per litre.

*Space for calculation*

________ : ________

lead content in roots lead content in shoots **1**

(*d*) The lead content is expressed in mg of lead per gram of dry mass of the plant. Explain the advantage of using dry mass rather than fresh mass.

________________________________

________________________________

________________________________ **1**

(*e*) Using the **Table** and **Bar Chart**, calculate the average lead content of the roots of the plants grown in the solution containing 30 mg of lead per litre.

*Space for calculation*

________________ mg lead **1**

(*f*) Explain why the presence of lead ions in the cells of French bean plants resulted in a decrease in growth.

________________________________

________________________________ **2**

DO NOT WRITE IN THIS MARGIN

*Marks*

**11.** (*a*) The graph shows the changes in body mass and height of a human male from the age of 1 to 21 years.

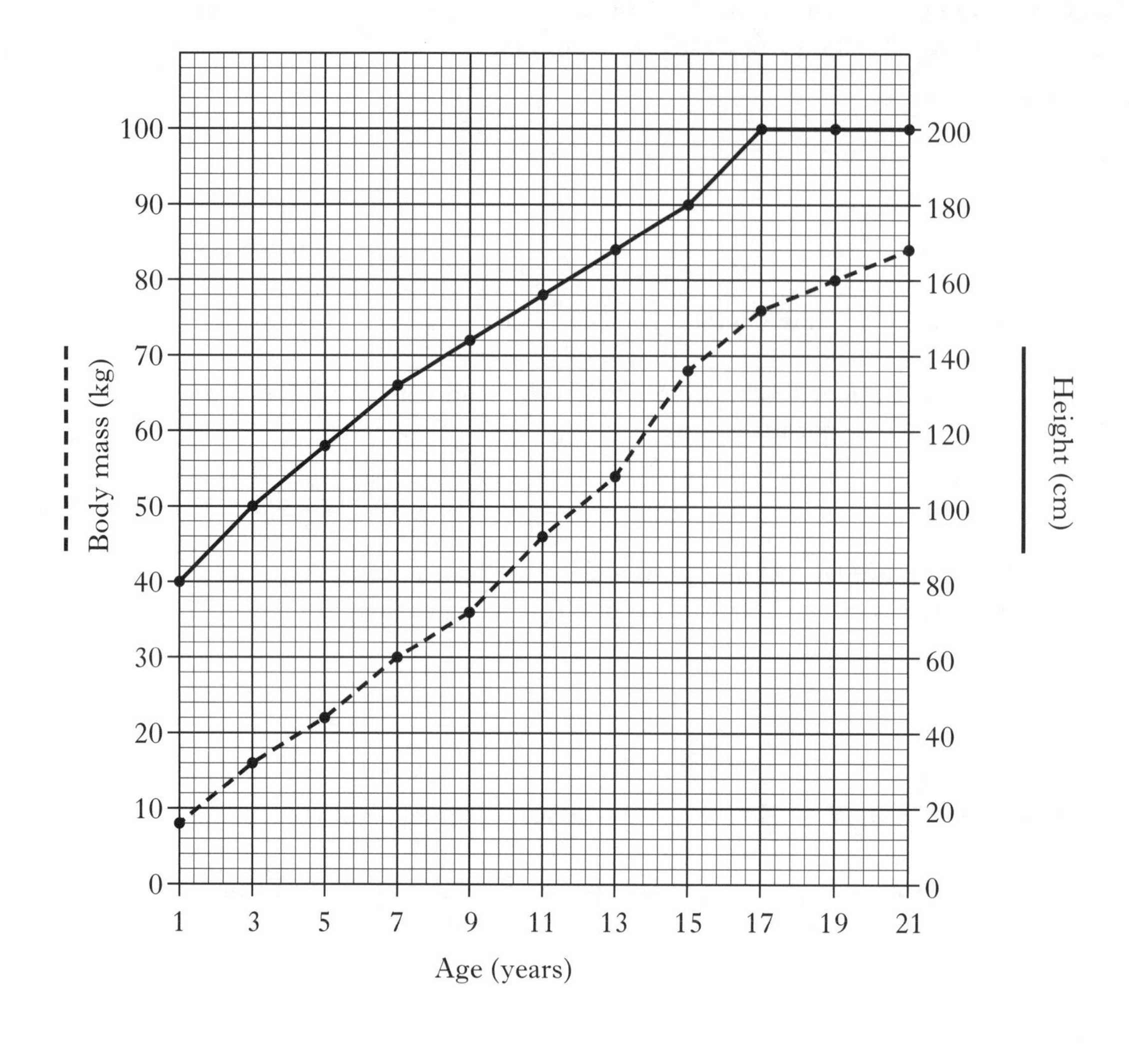


(i) Calculate the average yearly increase in body mass between 11 and 15 years.

*Space for calculation*

_____________ kg **1**

(ii) Tick (✓) the box to show the 4 year period in which the greatest increase in height occurred.

☐ 1–5 years ☐ 5–9 years ☐ 9–13 years ☐ 13–17 years **1**

DO NOT WRITE IN THIS MARGIN

**11. (continued)**

*Marks*

(*b*) The diagram shows how the pituitary gland is involved in the control of growth and development in humans.

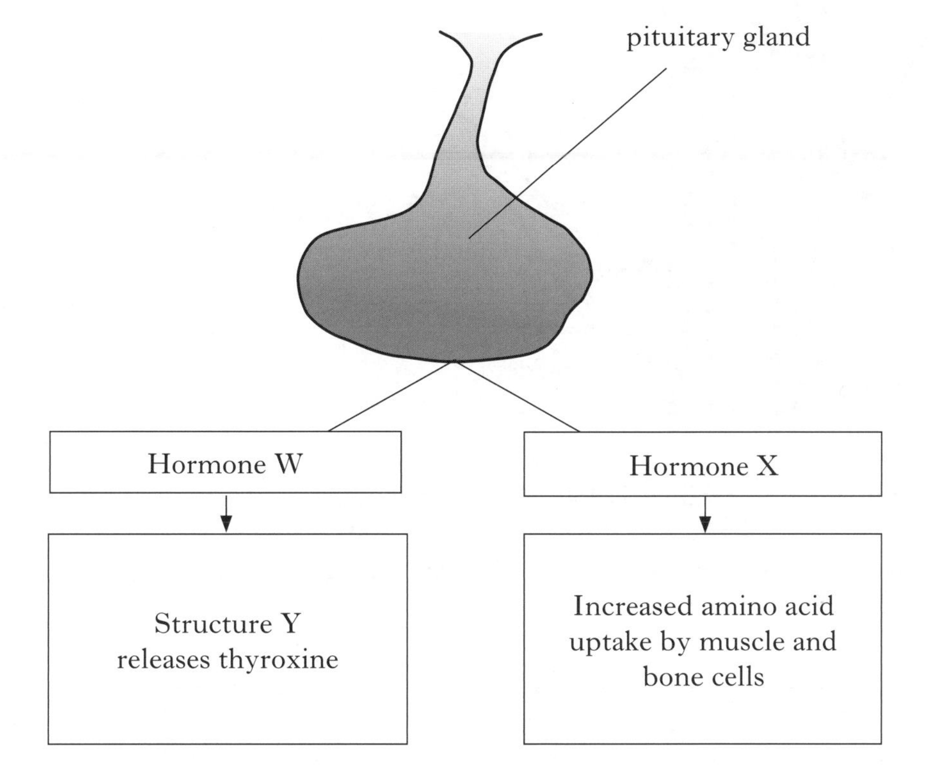


(i) Identify hormones W and X and structure Y.

Hormone W ______________________

Hormone X ______________________

Structure Y ______________________ **2**

(ii) Describe the effect of an increase in thyroxine production in humans.

______________________________________

______________________________________ **1**

**[Turn over**

DO NOT WRITE IN THIS MARGIN

*Marks*

**12.** (*a*) The diagram shows a vertical section through a shoot.

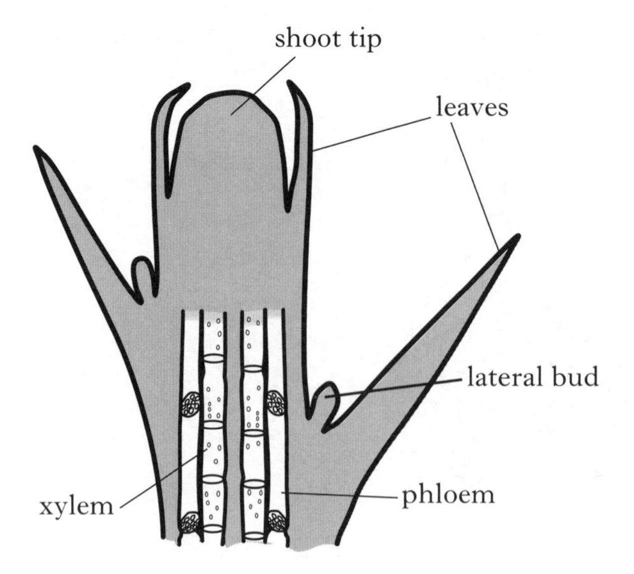


(i) Cells in the shoot tip produce indole acetic acid (IAA).

1 Describe how the IAA affects cellular activity resulting in an increase in shoot length.

______________________________________________

______________________________________________ **1**

2 Growth of lateral buds is inhibited by IAA.

State the term which describes this effect.

______________________________________________ **1**

(ii) Phloem and xylem are produced by the differentiation of unspecialised cells.

State how the differentiation of cells can be controlled by gene activity.

______________________________________________

______________________________________________

______________________________________________ **1**

(*b*) Macro-elements are important in the growth of plants.

(i) State the importance of magnesium in the growth of plants.

______________________________________________ **1**

(ii) Deficiency of phosphorus reduces overall growth of plants.

Give **one** other symptom of the deficiency of phosphorus in plants.

______________________________________________ **1**

DO NOT WRITE IN THIS MARGIN

*Marks*

**13.** The diagram shows an outline of the control of body temperature in a mammal.

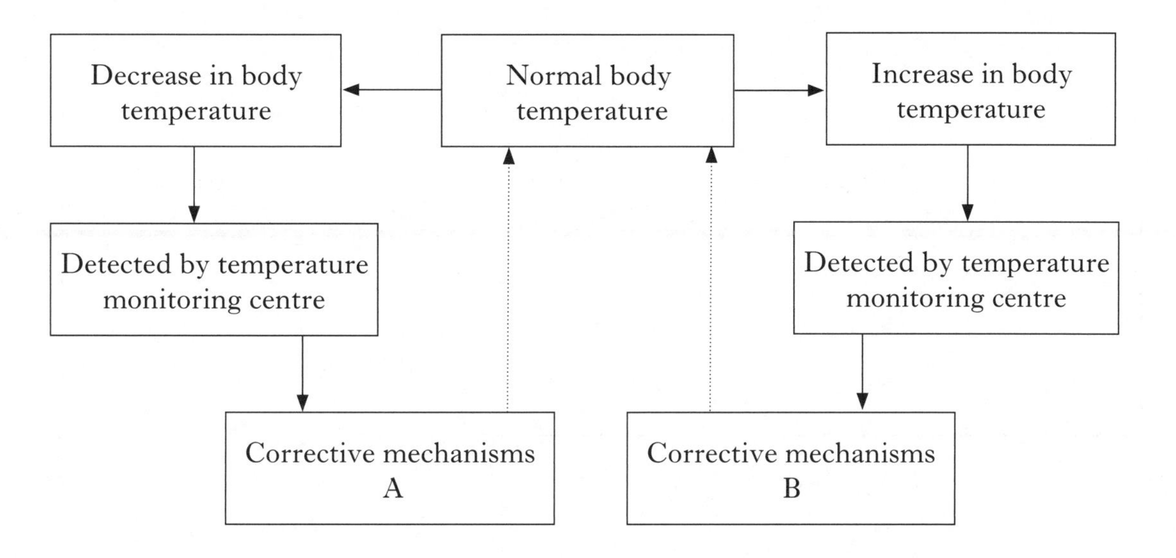


(*a*) (i) State the exact location of the temperature monitoring centre.

______________________________ **1**

(ii) The skin has effectors which are involved in corrective mechanisms A and B.

State how messages are sent from the temperature monitoring centre to the skin.

________________________________________________________ **1**

(iii) Give **one** example of a corrective mechanism B and explain how it would return the body temperature to normal.

Example ________________________________________________ **1**

Explanation ____________________________________________

________________________________________________________

________________________________________________________ **1**

(iv) Explain why maintaining body temperature within tolerable limits is important to the metabolism of mammals.

________________________________________________________

________________________________________________________ **1**

(*b*) Mammals obtain most of their heat from their metabolism.

Give the term which describes animals that obtain most of their body heat from their surroundings.

__________________________ **1**

DO NOT WRITE IN THIS MARGIN

*Marks*

**14.** (*a*) The diagram shows the relationship between a predator population and the population of its prey over a period of time.

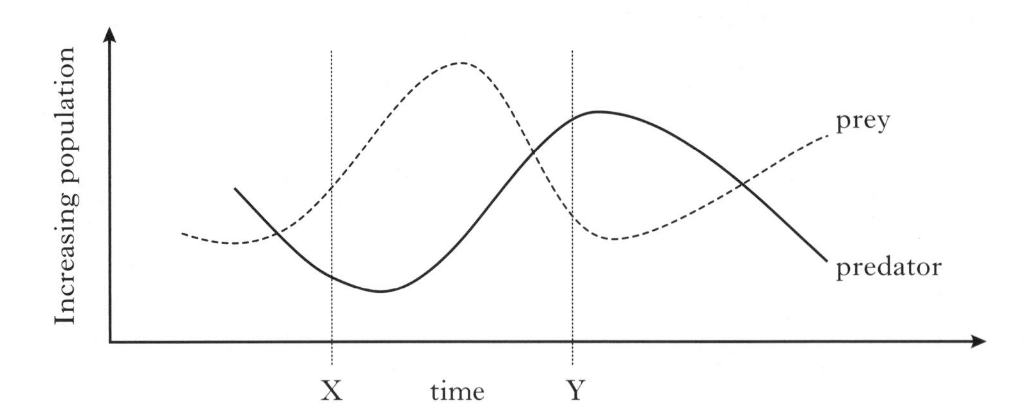


(i) Explain the changes in the population of **prey** between X and Y.

______________________________

______________________________

______________________________ **2**

(ii) Predation is a density-dependent factor.

Give **one** other density-dependent factor which influences animal populations.

______________________________ **1**

(*b*) Animal populations are monitored to provide data for a wide variety of purposes.

Complete the table to show the categories of species monitored and the use of the data collected.

| *Category of species* | *Use of data collected* |
|---|---|
| | ensure future supply for human use |
| pest | |
| | assess levels of pollution |
| endangered | |

**2**

DO NOT WRITE IN THIS MARGIN

*Marks*

**15.** (*a*) Flowering in some species of plant is affected by the periods of light and dark to which the plants are exposed.

The diagram below shows how flowering in plant species **P** and **Q** is affected by changing the periods of light and dark in 24 hours.

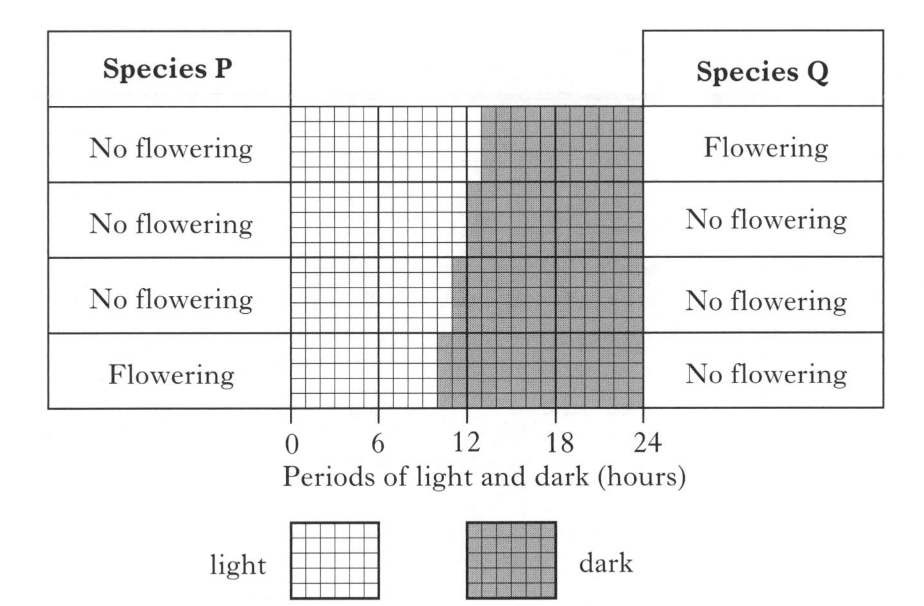


(i) Underline one alternative in each pair to make the sentence below correct.

Species **Q** is a {long / short} day plant which requires a critical dark period of less than {eleven / twelve} hours to flower. **1**

(ii) Predict the effect on the flowering of species **P** when exposed to the periods of light and dark shown below.

Justify your answer.

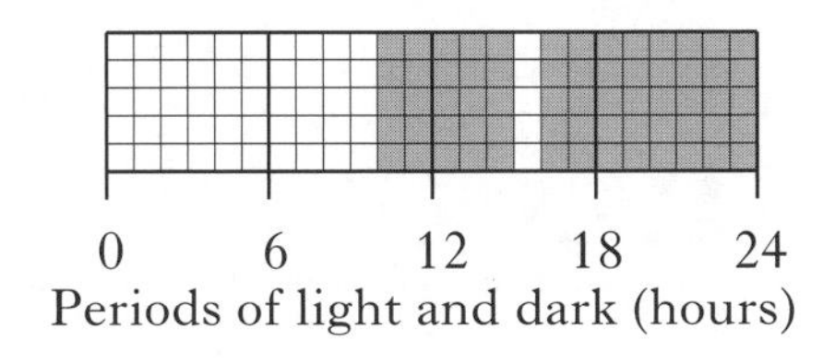


Prediction ______________________________ **1**

Justification ______________________________

______________________________ **1**

**15. (continued)** *Marks*

(*b*) (i) Ferrets are long day breeders and give birth six weeks after mating. Tick (✓) the box to show the season in which ferrets mate and explain how the timing of their breeding gives their offspring the best chance of survival.

Mating season Spring ☐ Autumn ☐

Explanation ______________________________________

______________________________________ **1**

(ii) What general term is used to describe the effect of light on the timing of breeding in mammals such as ferrets?

______________________________________ **1**

DO NOT WRITE IN THIS MARGIN

*Marks*

## SECTION C

**Both questions in this section should be attempted.**

Note that each question contains a choice.

**Questions 1 and 2 should be attempted on the blank pages which follow.**

**Supplementary sheets, if required, may be obtained from the Invigilator.**

**All answers must be written clearly and legibly in ink.**

**Labelled diagrams may be used where appropriate.**

**1.** Answer **either** A **or** B.

**A.** Write notes on the evolution of new species under the following headings:

(i) the role of isolation and mutation; **6**

(ii) natural selection. **4**

**(10)**

**OR**

**B.** Write notes on adaptations for obtaining food in animals under the following headings:

(i) the economics of foraging behaviour; **2**

(ii) cooperative hunting, dominance hierarchy and territorial behaviour. **8**

**(10)**

**In question 2, ONE mark is available for coherence and ONE mark is available for relevance.**

**2.** Answer **either** A **or** B.

**A.** Give an account of the structure of a mitochondrion and the role of the cytochrome system in respiration. **(10)**

**OR**

**B.** Give an account of phagocytosis and the role of lymphocytes in cellular defence. **(10)**

[*END OF QUESTION PAPER*]

**SPACE FOR ANSWERS**

DO NOT WRITE IN THIS MARGIN

SPACE FOR ANSWERS

DO NOT WRITE IN THIS MARGIN

**SPACE FOR ANSWERS**

DO NOT WRITE IN THIS MARGIN

**SPACE FOR ANSWERS**

SPACE FOR ANSWERS

DO NOT WRITE IN THIS MARGIN

**SPACE FOR ANSWERS**

DO NOT WRITE IN THIS MARGIN

**SPACE FOR ANSWERS**

ADDITIONAL GRAPH PAPER FOR QUESTION 4(*d*)

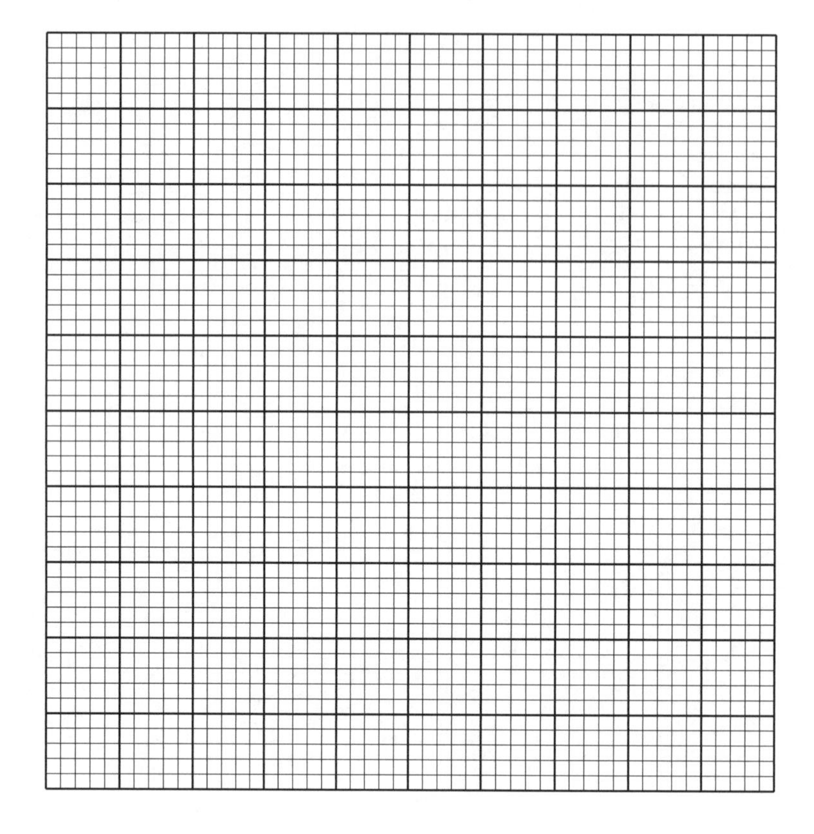

HIGHER

# 2013

[BLANK PAGE]

FOR OFFICIAL USE

| | | | | | |
|---|---|---|---|---|---|
| | | | | | |

Total for Sections B and C

# X007/12/02

NATIONAL QUALIFICATIONS 2013

WEDNESDAY, 15 MAY 1.00 PM – 3.30 PM

# BIOLOGY HIGHER

**Fill in these boxes and read what is printed below.**

Full name of centre

Town

Forename(s)

Surname

Date of birth

Day Month Year

Scottish candidate number

Number of seat

**SECTION A—Questions 1—30 (30 Marks)**

Instructions for completion of Section A are given on *Page two.*

For this section of the examination you must use an **HB pencil**.

**SECTIONS B AND C (100 Marks)**

1 (a) All questions should be attempted.

(b) It should be noted that in **Section C** questions 1 and 2 each contain a choice.

2 The questions may be answered in any order but all answers are to be written in the spaces provided in this answer book, **and must be written clearly and legibly in ink**.

3 Additional space for answers will be found at the end of the book. If further space is required, supplementary sheets may be obtained from the Invigilator and should be inserted inside the **front** cover of this book.

4 The numbers of questions must be clearly inserted with any answers written in the additional space.

5 Rough work, if any should be necessary, should be written in this book and then scored through when the fair copy has been written. If further space is required, a supplementary sheet for rough work may be obtained from the Invigilator.

6 Before leaving the examination room you must give this book to the Invigilator. If you do not, you may lose all the marks for this paper.

**Read carefully**

1 Check that the answer sheet provided is for **Biology Higher (Section A)**.

2 For this section of the examination you must use an **HB pencil**, and where necessary, an eraser.

3 Check that the answer sheet you have been given has **your name**, **date of birth**, **SCN** (Scottish Candidate Number) and **Centre Name** printed on it.

Do not change any of these details.

4 If any of this information is wrong, tell the Invigilator immediately.

5 If this information is correct, **print** your name and seat number in the boxes provided.

6 The answer to each question is **either** A, B, C or D. Decide what your answer is, then, using your pencil, put a horizontal line in the space provided (see sample question below).

7 There is **only one correct** answer to each question.

8 Any rough working should be done on the question paper or the rough working sheet, **not** on your answer sheet.

9 At the end of the examination, put the **answer sheet for Section A inside the front cover of this answer book**.

**Sample Question**

The apparatus used to determine the energy stored in a foodstuff is a

A calorimeter

B respirometer

C klinostat

D gas burette.

The correct answer is **A**—calorimeter. The answer **A** has been clearly marked in **pencil** with a horizontal line (see below).

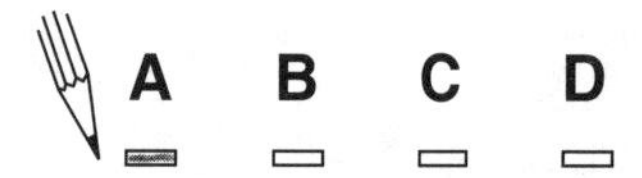


**Changing an answer**

If you decide to change your answer, carefully erase your first answer and using your pencil fill in the answer you want. The answer below has been changed to **D**.

A B C D

## SECTION A

**All questions in this section should be attempted.**

**Answers should be given on the separate answer sheet provided.**

**1.** Plant cell walls are composed mainly of

A cellulose

B phospholipid

C collagen

D starch.

**2.** Visking tubing is selectively permeable. In the experiment shown below to demonstrate osmosis, the following results were obtained.

Initial mass of Visking tubing + contents = 10·0 g

Mass of Visking tubing + contents after experiment = 11·8 g

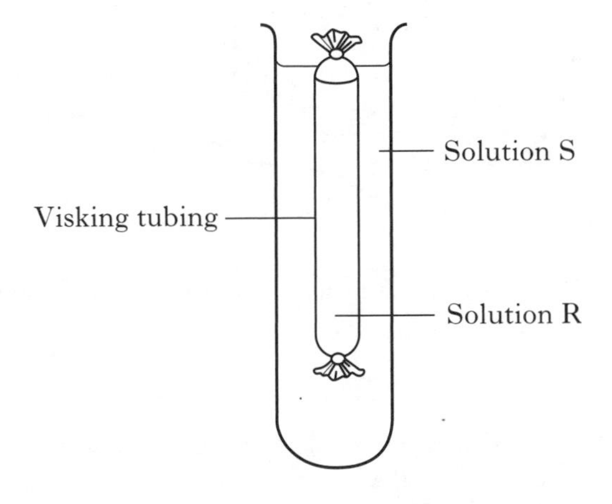


The results shown could be obtained when

A R is a 5% salt solution and S is a 10% salt solution

B R is a 10% salt solution and S is a 5% salt solution

C R is water and S is a 10% salt solution

D R is water and S is a 5% salt solution.

**3.** The diagram below represents part of the Calvin cycle within a chloroplast.

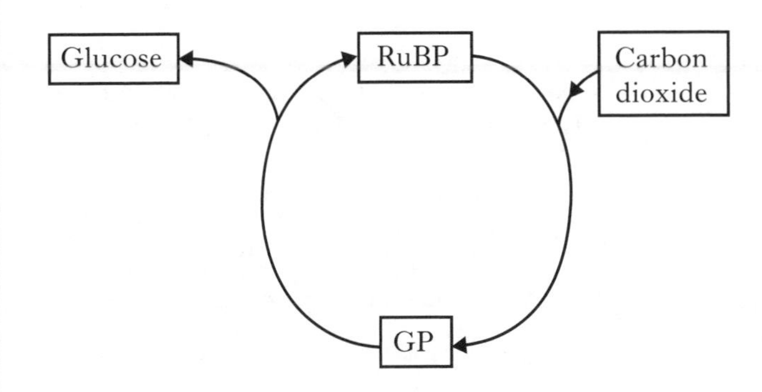


Which line in the table below shows the effect of decreasing $CO_2$ availability on the concentrations of RuBP and GP in the cycle?

| | *RuBP concentration* | *GP concentration* |
|---|---|---|
| A | decrease | decrease |
| B | increase | increase |
| C | decrease | increase |
| D | increase | decrease |

**[Turn over**

4. Which of the following must be present in a living cell for glycolysis to occur?

A Glucose and ATP

B Pyruvic acid and oxygen

C Glucose and oxygen

D Pyruvic acid and ATP

5. Which of the following chemical changes in cells results in the synthesis of most ATP?

A Glucose to pyruvic acid

B Pyruvic acid to lactic acid

C Pyruvic acid to acetyl group

D Pyruvic acid to carbon dioxide and water

6. The graph below shows the effect of the carbon dioxide concentration of inhaled air on the breathing rate of an individual.

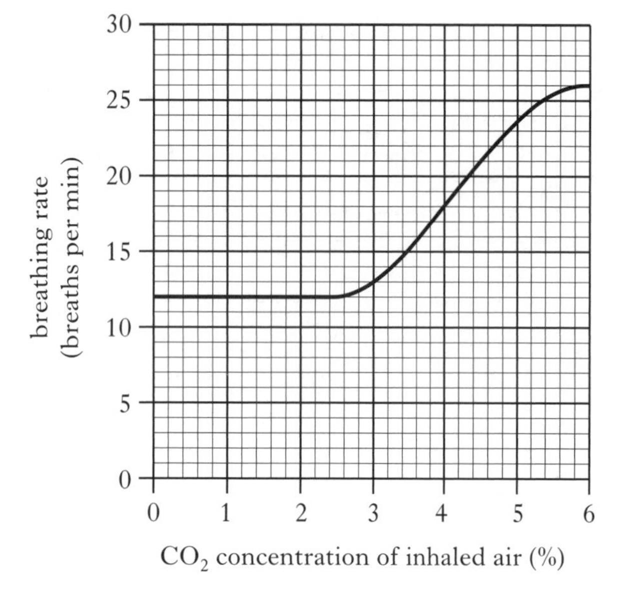


If the volume of one breath is 0·5 litre, what volume of air will be breathed in one minute when the $CO_2$ concentration is 4%?

A 6 litres

B 9 litres

C 18 litres

D 36 litres

7. If ten percent of the bases in a molecule of DNA are adenine, what is the ratio of adenine to guanine in the same molecule?

A 1:1

B 1:2

C 1:3

D 1:4

8. The diagram below shows some structures within an animal cell.

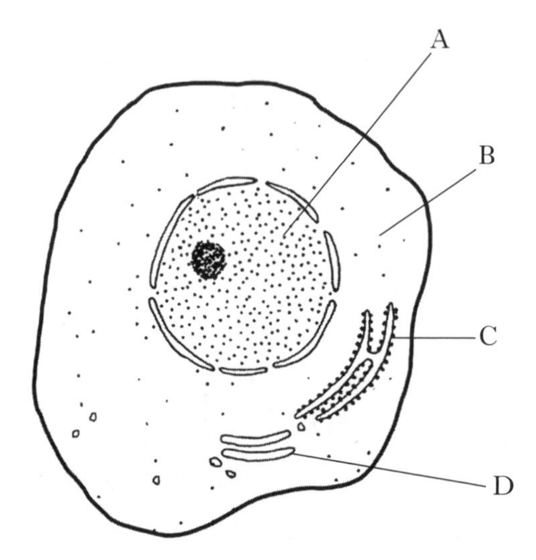


Where does synthesis of messenger RNA take place?

9. The processing of proteins prior to secretion takes place in the

A endoplasmic reticulum

B Golgi apparatus

C ribosomes

D vesicles.

10. Phagocytes contain lysosomes that

A recognise foreign antigens on bacteria

B produce antibodies to destroy viruses

C surround and engulf invading viruses

D contain enzymes which destroy bacteria.

**11.** In garden pea plants the genes for height and flower colour are on different chromosomes. The allele **T** (tall) is dominant to the allele **t** (dwarf) and the allele **R** (purple flowers) is dominant to the allele **r** (white flowers).

A true-breeding tall plant with purple flowers was crossed with a dwarf plant with white flowers.

The $F_1$ generation were self-pollinated and 64 plants were obtained in the $F_2$ generation.

How many of the $F_2$ generation would be expected to be tall with white flowers?

A 36

B 16

C 12

D 4

**12.** Which line in the table correctly shows characteristics of mutant alleles?

| | *Frequency* | *Occurrence* |
|---|---|---|
| A | high | random |
| B | high | non-random |
| C | low | random |
| D | low | non-random |

**13.** Which of the following has occurred as a result of natural selection?

A Modern varieties of potato have been produced from wild varieties.

B Ayrshire cows have been bred to increase milk yield.

C Bacteria have developed resistance to some antibiotics.

D Some tomato plants produced by somatic fusion have resistance to fungal diseases.

**14.** The dark variety of the peppered moth became common in industrial areas of Britain following the increase in production of soot during the Industrial Revolution.

The increase in the dark form was due to

A dark moths migrating to areas which offered the best camouflage

B a change in the prey species taken by birds

C an increase in the mutation rate

D a change in selection pressure.

**15.** An eel was transferred from salt water to fresh water.

The table shows how the drinking rate of the eel changed in the six hour period after transfer.

| *Time after transfer* (hours) | *Drinking rate* ($cm^3$ per kg body mass per hour) |
|---|---|
| 0 | 20 |
| 1 | 17 |
| 2 | 15 |
| 3 | 12 |
| 4 | 10 |
| 5 | 8 |
| 6 | 2 |

What is the average hourly decrease in drinking rate over the six hour period after transfer?

A 3 $cm^3$ per kg body mass per hour

B 12 $cm^3$ per kg body mass per hour

C 14 $cm^3$ per kg body mass per hour

D 18 $cm^3$ per kg body mass per hour

**[Turn over**

**16.** Which line in the table below describes the action of chloride secretory cells in the gills and the glomerular filtration rate of a salmon living in sea water?

| | *Action of chloride secretory cells in the gills* | *Glomerular filtration rate* |
|---|---|---|
| A | excretes salts | high |
| B | excretes salts | low |
| C | absorbs salts | low |
| D | absorbs salts | high |

**17.** The diagram below shows a stoma and its guard cells in the lower epidermis of a leaf.

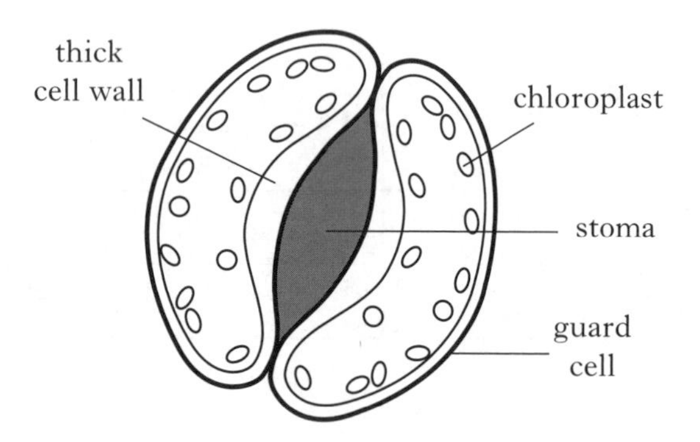


Which line of the table below describes the conditions which result in the stoma being fully open as shown in the diagram?

| | *Lighting conditions* | *Condition of guard cells* |
|---|---|---|
| A | light | flaccid |
| B | dark | flaccid |
| C | light | turgid |
| D | dark | turgid |

**18.** The Soft Brome grass and the Storksbill are species of plant which grow in the grasslands of California. The Storksbill has a more extensive root system, but does not grow as tall as the Soft Brome grass.

From this information, in which of the following conditions would the Storksbill be expected to survive better than Soft Brome grass?

A Drought

B High soil moisture levels

C High light intensity

D Shade

**19.** Two plants of different species had their carbon dioxide ($CO_2$) uptake and output measured in relation to light intensity. The results are shown below.

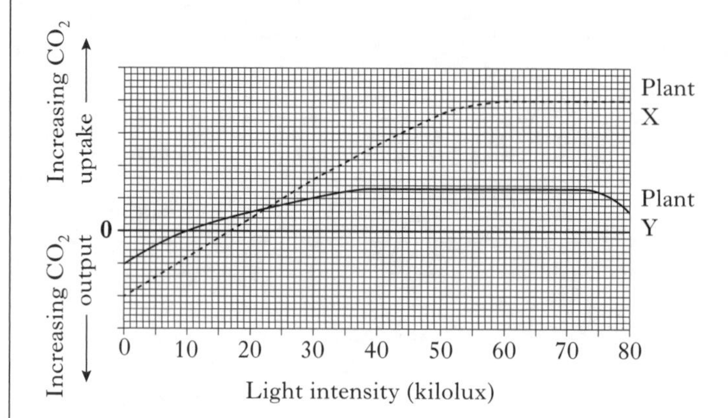


Which line in the table below is a correct interpretation of the graph?

| | *Type of plant* | *Light intensity at which compensation point is reached* |
|---|---|---|
| A | X is a shade plant | 17 kilolux |
| B | X is a sun plant | 60 kilolux |
| C | Y is a shade plant | 10 kilolux |
| D | Y is a sun plant | 40 kilolux |

**20.** A substitution mutation in a gene results in a triplet of bases TTC being changed to TCC. The amino acid lysine is coded for by TTC and arginine is coded for by TCC.

The effect of this mutation on the resulting protein would be that

A all lysine molecules would be replaced by arginine molecules throughout the protein

B one lysine molecule would replace arginine at one point in the protein

C all arginine molecules would be replaced by lysine molecules throughout the protein

D one arginine molecule would replace lysine at one point in the protein.

**21.** The diagram below shows a section through a young plant stem. In which region would a meristem be found?

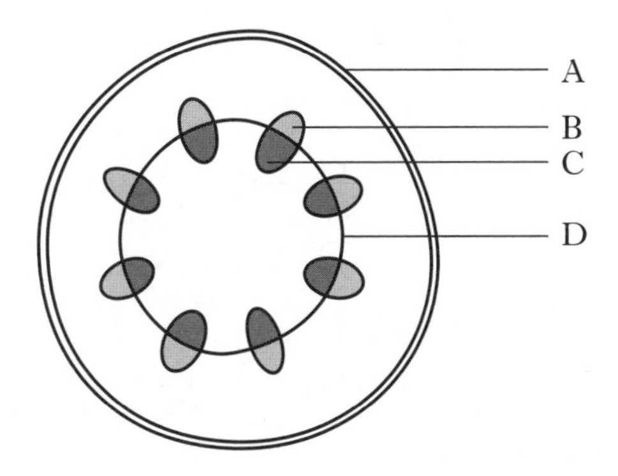


**22.** The ability of a plant to replace damaged parts through new growth is called

A abscission

B regeneration

C differentiation

D apical dominance.

**23.** Which of the following statements best defines the term population density?

A The number of individuals of a species present per unit area of a habitat.

B The number of individuals of all species present in a habitat.

C The maximum number of individuals of all species which the resources of a habitat can support.

D The maximum number of individuals of a species which the resources of a habitat can support.

**24.** The apparatus shown below was used to investigate the effects of phosphate on the growth of grass seedlings.

The experiment was repeated using different concentrations of phosphate and the height of the seedlings was recorded after 6 weeks of growth.

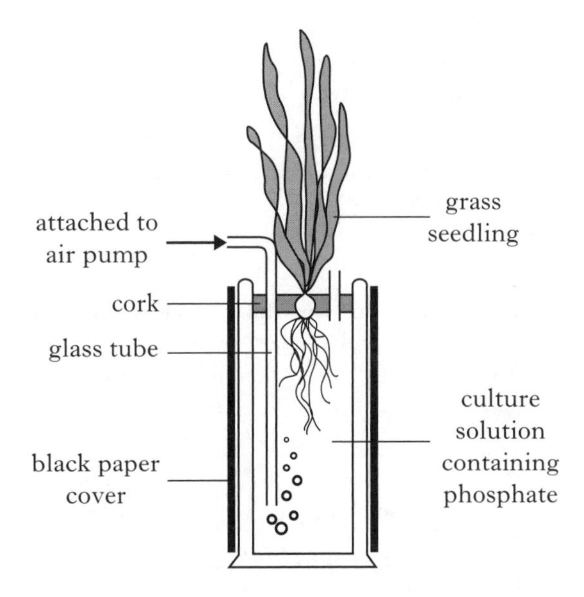


Two variables that must be kept the same are

A temperature and concentration of phosphate in the culture solution

B concentration of phosphate in the culture solution and light intensity

C light intensity and temperature

D temperature and the height of the seedlings.

25. Liver tissue contains an enzyme which breaks down alcohol. The graph below shows the effect of different concentrations of copper ions on the breakdown of alcohol by this enzyme over a 30 minute period.

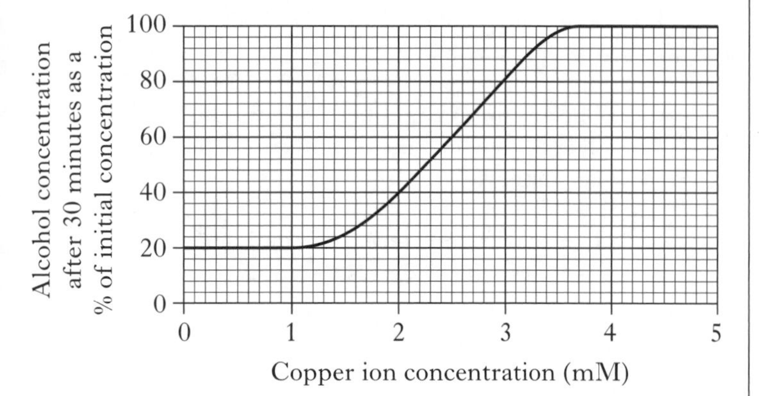


Which of the following conclusions can be drawn from the graph?

A 4·5 mM copper has no effect on enzyme activity.

B 2·5 mM copper halves the enzyme activity.

C 0·5 mM copper completely inhibits enzyme activity.

D Enzyme activity increases when copper concentration is increased from 1 mM to 2 mM.

26. Which line in the table below correctly identifies the substances required for normal blood-clotting and for the prevention of rickets?

| | *Normal blood-clotting* | *Prevention of rickets* |
|---|---|---|
| A | calcium | vitamin D |
| B | vitamin D | calcium |
| C | iron | vitamin D |
| D | calcium | iron |

27. Nicotine is a chemical which may affect fetal development.

The diagram shows the stages of development when major and minor malformations of organs may occur if there is exposure to nicotine during the first twelve weeks of pregnancy.

Key
major malformation
minor malformation

| *Organ* | *Stage of development* (weeks of pregnancy) | | | | | | | | | | | |
|---|---|---|---|---|---|---|---|---|---|---|---|---|
| | 1 | 2 | 3 | 4 | 5 | 6 | 7 | 8 | 9 | 10 | 11 | 12 |
| brain | | | major | major | major | minor | minor | minor | minor | minor | minor | minor |
| ear | | | | major | major | major | major | major | major | minor | minor | minor |
| limbs | | | | | major | major | major | minor | | | | |
| genitalia | | | | | | | major | major | major | minor | minor | minor |

For how many weeks during pregnancy is there a possibility of major malformations to organs during development?

A 6

B 7

C 9

D 14

28. A species of short-day plant only flowers if the number of hours of continuous darkness in its 24 hour cycle is at least at the critical value shown in chart below.

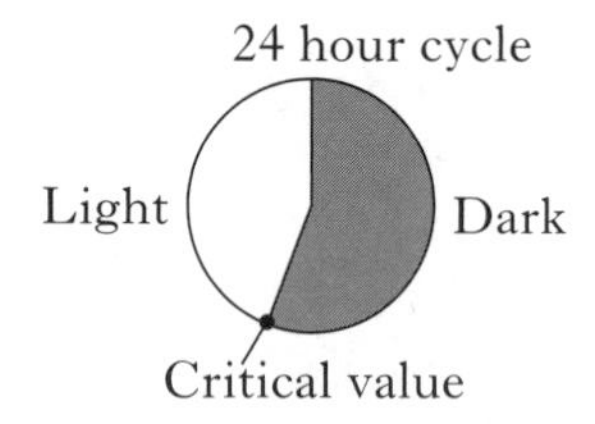


Three plants of this species were exposed to different photoperiods as shown below.

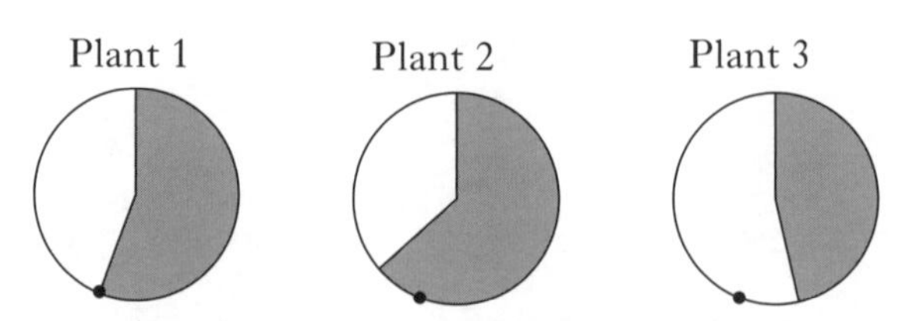


Which plant(s) would be expected to flower?

A 1 only

B 2 only

C 1 and 2 only

D 2 and 3 only

**29.** The diagram below shows a section through the skin of a mammal.

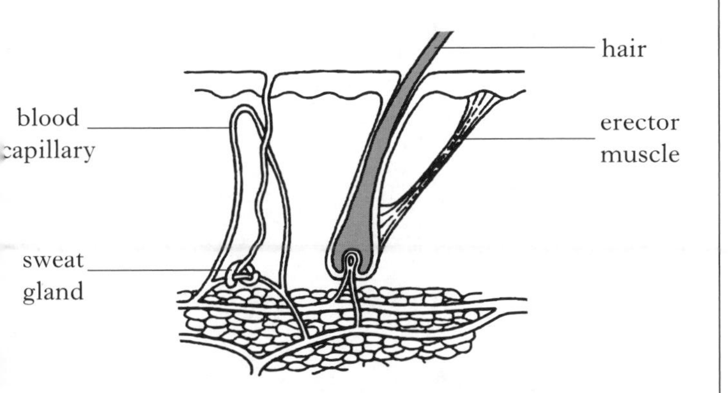


Which line in the table below correctly identifies the state of the erector muscle and the change in blood flow in the capillary which would be expected if the skin was exposed to low temperature?

| | *State of erector muscle* | *Change to blood flow in capillary* |
|---|---|---|
| A | contracted | increase |
| B | contracted | decrease |
| C | relaxed | increase |
| D | relaxed | decrease |

**30.** Which of the following factors influencing population change in an animal species is density-independent?

A Disease

B Food availability

C Temperature

D Predation

**Candidates are reminded that the answer sheet MUST be returned INSIDE the front cover of this answer book.**

**[Turn over for SECTION B on *Page ten***

DO NOT WRITE IN THIS MARGIN

## SECTION B

**All questions in this section should be attempted.**

**All answers must be written clearly and legibly in ink.**

*Marks*

**1.** (*a*) The diagram below shows part of a DNA molecule.

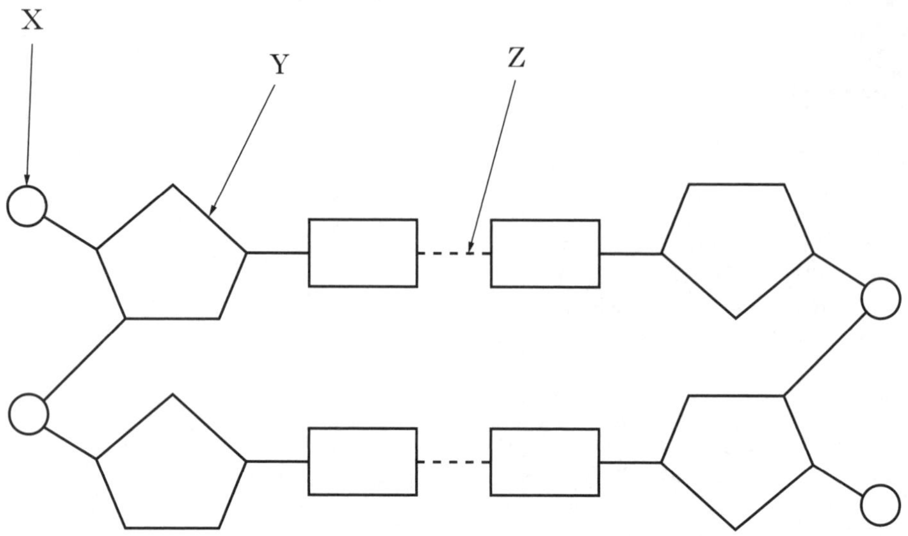

X

Y

Z

(i) Name components X and Y.

X ______________________

Y ______________________ **1**

(ii) Name the type of bond shown at Z.

______________________ **1**

DO NOT WRITE IN THIS MARGIN

*Marks*

**1. (continued)**

(*b*) (i) The flowchart below describes steps in the process of DNA replication.

Complete the boxes to describe what happens at **Step 2** and **Step 4**.

| | |
|---|---|
| **Step 1** | Original DNA double helix unwinds. |
| **Step 2** | |
| **Step 3** | Free DNA nucleotides bond with complementary nucleotides on the original DNA strands. |
| **Step 4** | |
| **Step 5** | Double strands twist and two new DNA double helices are formed. |

2

(ii) Other than the original DNA strand and free DNA nucleotides, give **one** substance needed for DNA replication.

______________________________ 1

(iii) State the importance of DNA replication to cells.

______________________________

______________________________

______________________________ 1

DO NOT WRITE IN THIS MARGIN

*Marks*

**2.** (*a*) Cherry tree leaves are attacked by greenfly. The leaves contain cyanogenic glycosides which are broken down to release cyanide when greenfly damage them. The cyanide acts as a defence against **most** greenfly species.

The graph below shows the average number of individuals of a species of greenfly per leaf and the concentration of cyanogenic glycosides in the leaves of a cherry tree over a 60 day period.

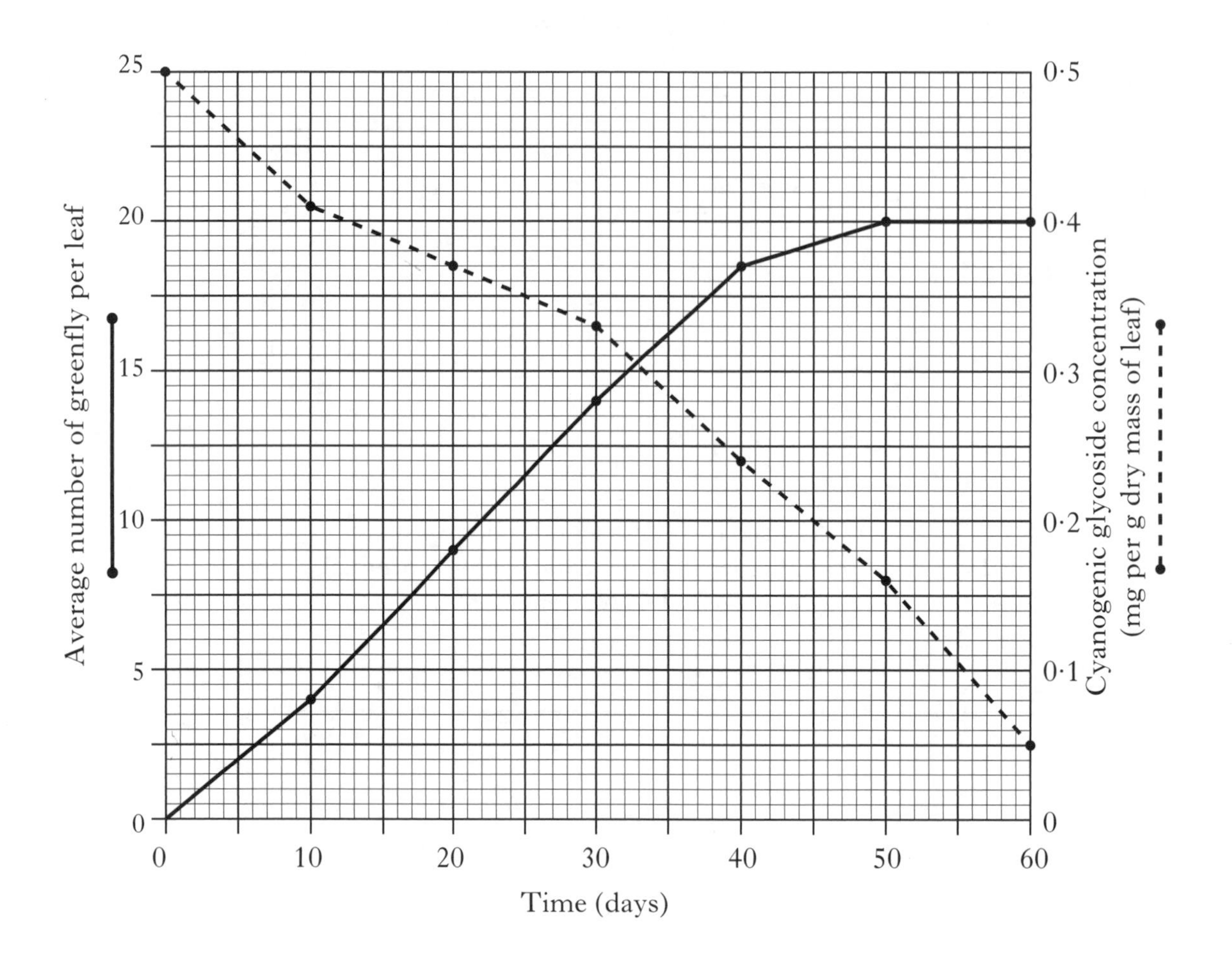


(i) Explain how changes in the population of greenfly account for the fall in the cyanogenic glycoside concentration throughout the period.

________________________________________

________________________________________

________________________________________ **2**

(ii) What evidence is there that this greenfly species is resistant to the effects of cyanide?

________________________________________ **1**

DO NOT WRITE IN THIS MARGIN

*Marks*

**2. (*a*) (continued)**

(iii) Calculate the average increase **per day** in the number of greenfly per leaf between day 10 and day 50.

*Space for calculation*

Average increase per day ____________________ **1**

(iv) State the cyanogenic glycoside concentration when the average number of greenfly per leaf was 14.

____________________ mg per gram dry mass of leaf **1**

(*b*) Some plants secrete sticky resin in response to damage.

Explain how resin protects the plants.

________________________________________

________________________________________ **1**

[Turn over

DO NOT WRITE IN THIS MARGIN

Marks

**3.** The diagram shows a cell from the root epidermis of the Spanish reed.

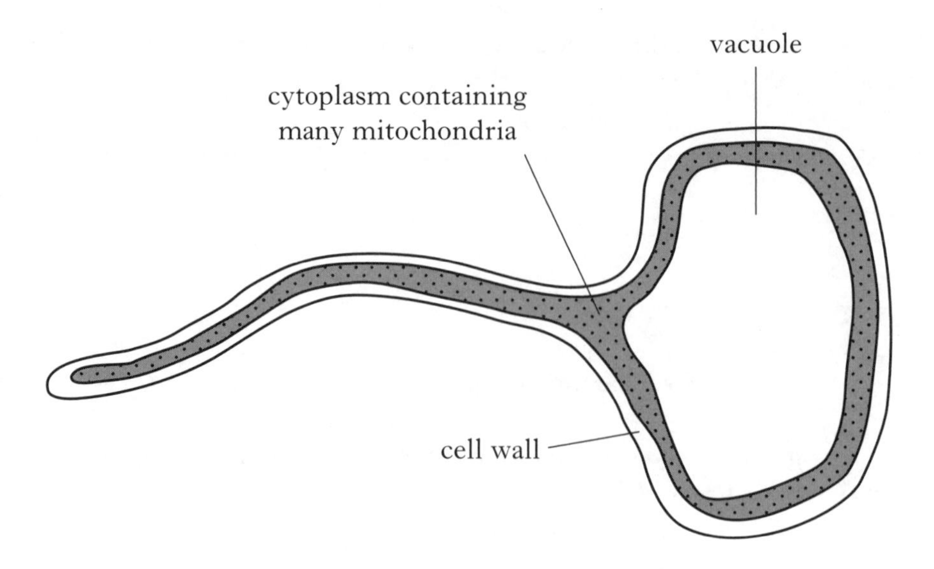


(*a*) Name this type of cell and describe how it is adapted for the absorption of water from the soil.

Type of cell ______________________________ **1**

Description ______________________________

______________________________ **1**

(*b*) The Spanish reed is adapted to grow in soil with high salt concentration. Salt enters the plant due to its high concentration in the soil. In order to survive, the plant must remove the excess salt from its cells.

Name the process by which the salt is removed and describe the role of mitochondria in this process.

Process ______________________________ **1**

Role of mitochondria in this process ______________________________

______________________________

______________________________ **1**

**[Turn over for Question 4 on *Page sixteen***

DO NOT WRITE IN THIS MARGIN

**4.** In an investigation into the effect of lead ion concentration on respiration in yeast, two flasks were set up as described below. *Marks*

| *Flask* | *Contents* |
|---|---|
| A | 200 $cm^3$ glucose solution + 5 $cm^3$ 0·2% lead nitrate solution |
| B | 200 $cm^3$ glucose solution + 5 $cm^3$ 1·0% lead nitrate solution |

The flasks were placed in a water bath at 20 °C for 10 minutes. After this time 2·5 $cm^3$ of yeast suspension was added to each and oxygen sensors fitted as shown in the diagram below.

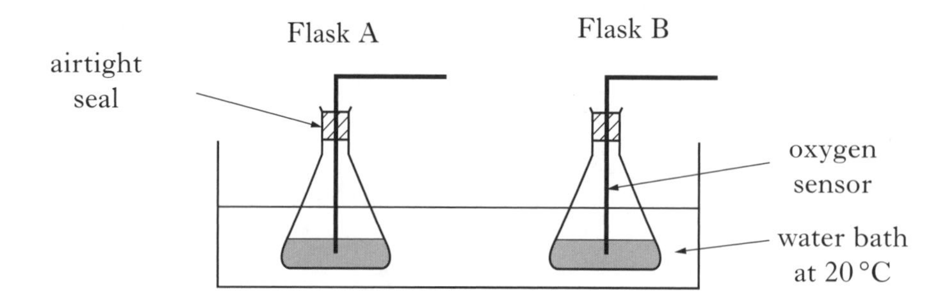


The flasks were left for a **further** 10 minutes and then oxygen concentration was measured in each flask every 20 seconds for 2 minutes.

The results are shown in the table below.

| *Time* (s) | *Oxygen concentration* (mg per litre) | |
|---|---|---|
| | *Flask A* *0·2% lead nitrate* | *Flask B* *1·0% lead nitrate* |
| 0 | 10·2 | 10·8 |
| 20 | 8·4 | 9·3 |
| 40 | 6·1 | 7·6 |
| 60 | 3·8 | 6·2 |
| 80 | 1·7 | 5·1 |
| 100 | 0·2 | 4·0 |
| 120 | 0·0 | 3·2 |

(*a*) (i) Identify **one** variable, not already mentioned, which would have to be kept constant so that valid conclusions could be drawn.

________________________________________ **1**

(ii) Explain why the flasks were left for 10 minutes **before** the yeast suspension was added.

________________________________________

________________________________________ **1**

DO NOT WRITE IN THIS MARGIN

**4. (a) (continued)**

*Marks*

(iii) Explain why the flasks were left for a **further** 10 minutes after the yeast suspensions were added before measurement of oxygen concentrations were taken.

______________________________________________

______________________________________________ **1**

(*b*) On the grid provided, draw a line graph to show the oxygen concentration in **Flask A** against time. Use an appropriate scale to fill most of the grid. (Additional graph paper, if required, will be found on Page forty.)

**2**

(*c*) Using information from the table, state the effect of increasing lead ion concentration on the aerobic respiration of yeast.

______________________________________________

______________________________________________ **1**

(*d*) Bubbles of gas appeared in both flasks throughout the investigation.

(i) Name this gas.

______________________________________________ **1**

(ii) Explain why this gas continued to be produced in **Flask A** at 120s.

______________________________________________

______________________________________________ **1**

*Marks*

**5.** (*a*) The diagram below shows two pairs of homologous chromosomes from a cell dividing by meiosis in a flowering plant. The letters represent alleles.

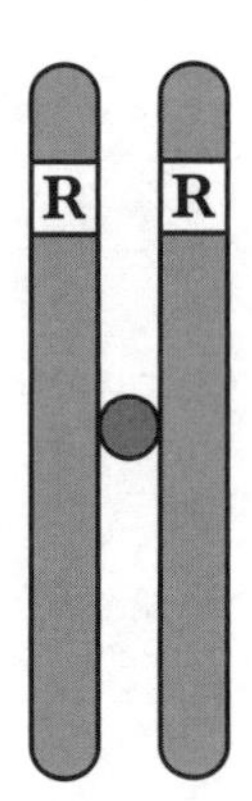

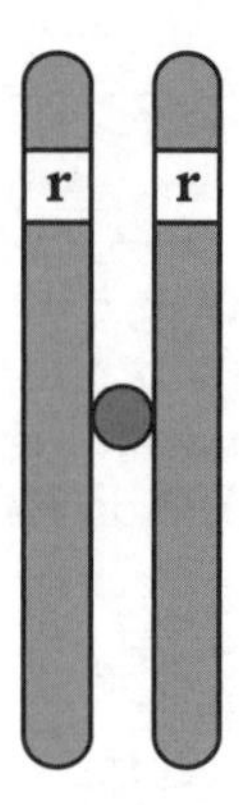

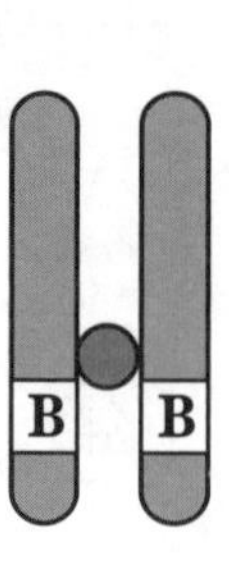

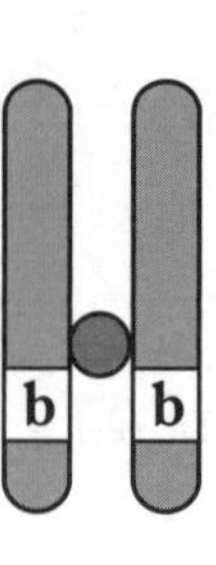


(i) Use letters from the diagram to give the genotypes of all four possible gametes which could be produced by this plant.

__________ __________ __________ __________ **1**

(ii) The diagram below shows the same pairs of homologous chromosomes separating during the first meiotic division.

The position of **one** of the alleles is shown.

Complete the diagram by adding letters to the remaining boxes to show the positions of the alleles that would result in the production of a gamete with the genotype **Rb**.

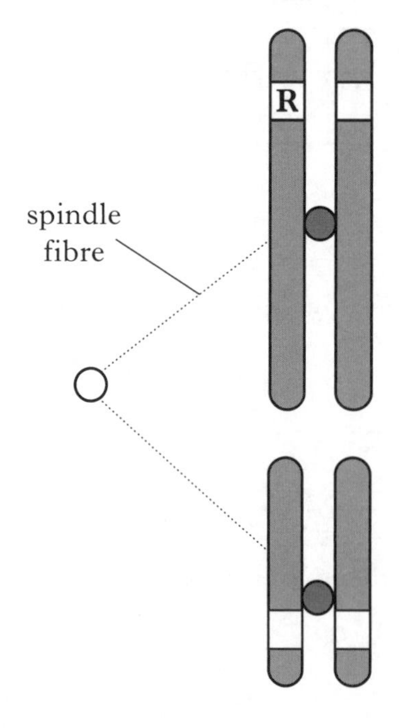

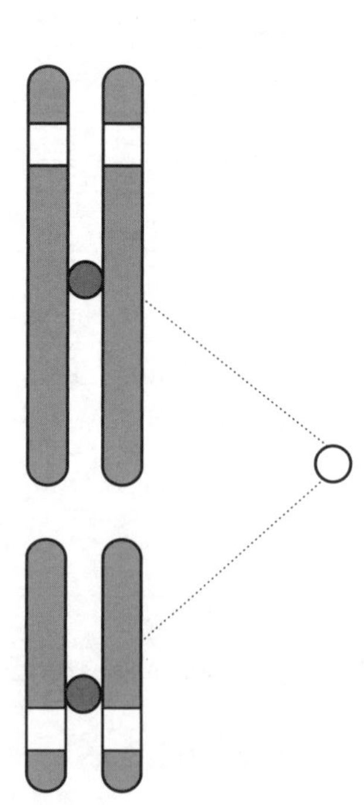

**1**

DO NOT WRITE IN THIS MARGIN

*Marks*

**5. (*a*) (continued)**

(iii) The chromosomes shown have two copies of each gene.

Describe what occurs during the second meiotic division which results in every gamete produced having only one copy of each gene.

________________________________________

________________________________________ **1**

(*b*) The diagram below shows another pair of homologous chromosomes from the same plant cell.

The letters for the alleles of two linked genes are shown.

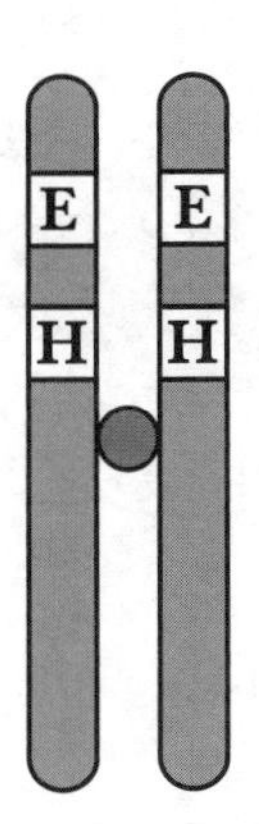


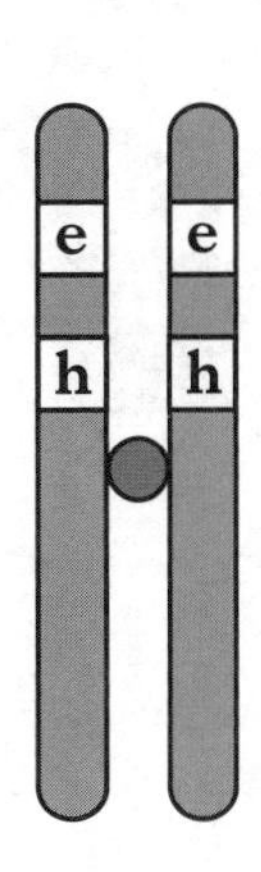


(i) The grid below shows the genotypes of the possible gametes produced by this cell following meiosis.

Complete the grid by ticking (✓) the boxes to show which of the gametes produced are recombinants.

| *Genotype of gametes* | EH | Eh | eH | eh |
|---|---|---|---|---|
| *Recombinant gametes* | | | | |

**1**

(ii) Name the process that occurs during meiosis that results in the production of recombinant gametes.

State the importance of this process to the plant species.

Process ________________________________ **1**

Importance ______________________________

________________________________________ **1**

DO NOT WRITE IN THIS MARGIN

*Marks*

**6.** (*a*) The map below shows the locations of six populations of the house mouse on the island of Madeira.

Studies on the mice have shown that speciation is occurring.

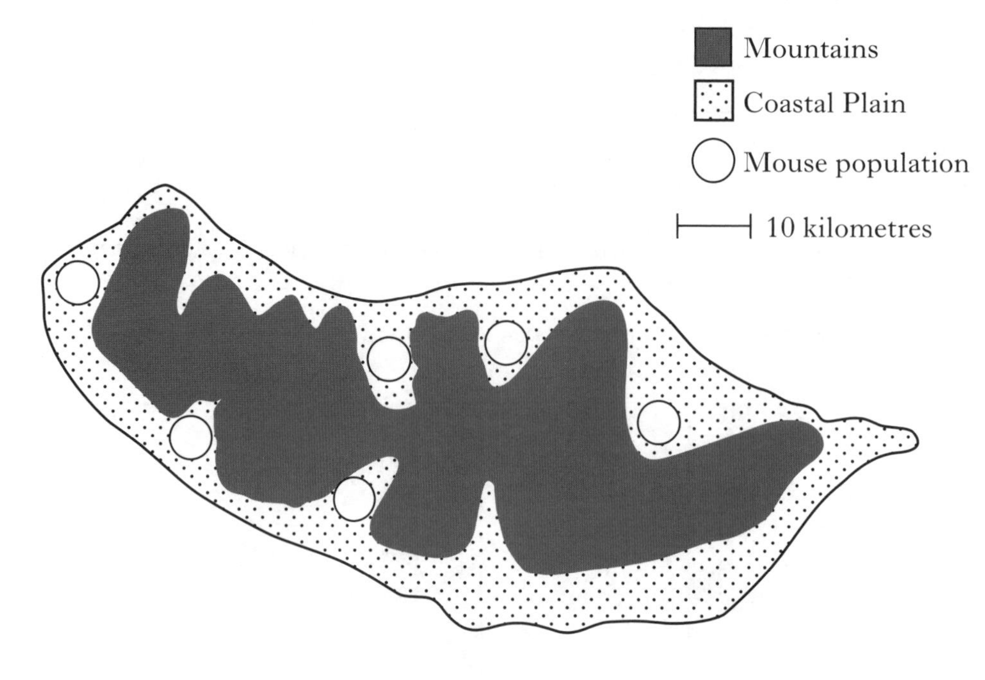


(i) Using information in the diagram, name the isolating mechanism involved in speciation of the mice.

______________________________ **1**

(ii) Explain the importance of isolating mechanisms in the evolution of a new species.

______________________________

______________________________ **1**

(iii) Describe evidence which would confirm that the populations of mice had evolved to become separate species.

______________________________

______________________________ **1**

DO NOT WRITE IN THIS MARGIN

*Marks*

**6. (continued)**

(*b*) **Diagram A** shows a bacterial plasmid containing genes for resistance to the antibiotics tetracycline and ampicillin.

**Diagram B** shows this plasmid after it had been genetically engineered by inserting a human blood-clotting gene.

**Diagram A**

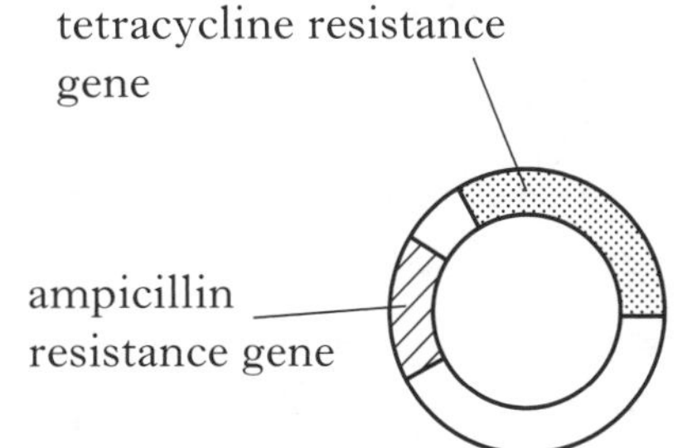


**Diagram B**

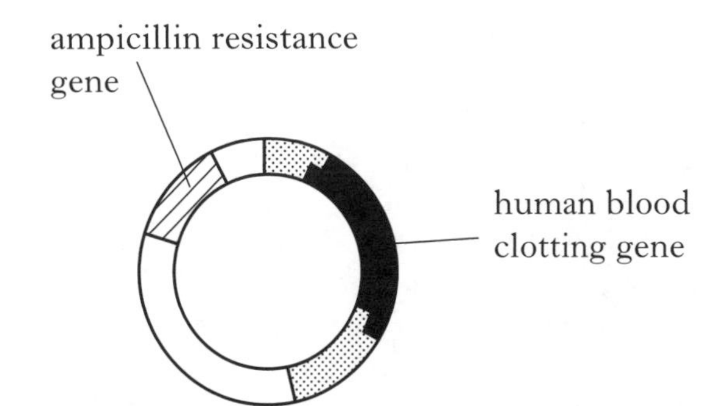


(i) Give a technique which could be used to locate the position of the blood-clotting gene on a human chromosome.

________________________________________ **1**

(ii) Two different enzymes are used to produce the genetically engineered plasmid.

Complete the table to show the function of each enzyme.

| *Enzyme* | *Function* |
|---|---|
| endonuclease | |
| ligase | |

**2**

(iii) The genetically engineered plasmids were inserted into bacteria.

Using information from the diagrams, explain why these bacteria were **not** resistant to tetracycline.

________________________________________

________________________________________

________________________________________ **2**

DO NOT WRITE IN THIS MARGIN

*Marks*

**7.** Red-green colour deficiency in humans is caused by a mutation in the gene coding for one of the proteins needed for normal colour vision.

This gene is sex-linked and the allele for colour deficiency **d** is recessive to the allele for normal colour vision **D**.

The diagram below shows inheritance of red-green colour deficiency in a family.

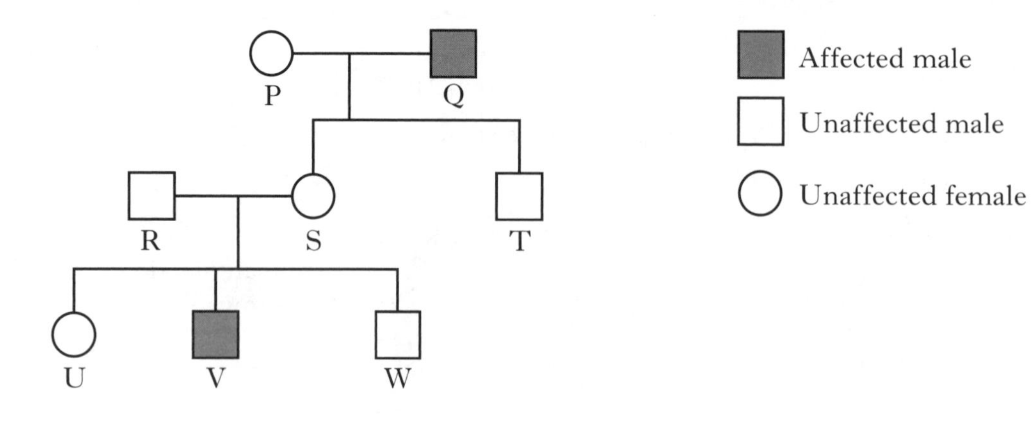


(*a*) Give the genotype of each of the following individuals.

Q ________________

S ________________

W ________________ **2**

(*b*) Explain how information from the diagram confirms that the allele causing red-green colour deficiency is recessive.

____________________________________________________

____________________________________________________

____________________________________________________ **1**

(*c*) Explain why males are more likely to be affected by red-green colour deficiency than females.

____________________________________________________

____________________________________________________

____________________________________________________ **1**

DO NOT WRITE IN THIS MARGIN

*Marks*

**8.** (*a*) Marram grass is adapted to reduce water loss. Its leaves contain hinge cells which let them curl when the soil in which the plant grows is dry.

When the soil is moist, the hinge cells make the leaves uncurl.

The diagram below shows a section through a **curled** leaf of marram grass.

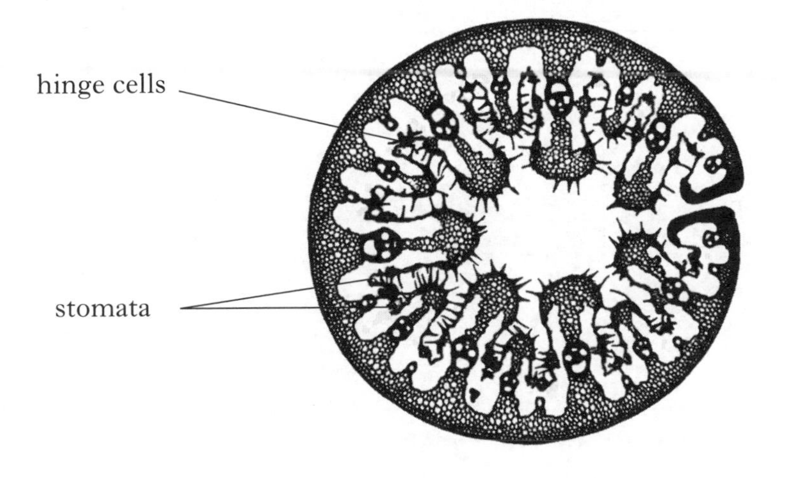


(i) Explain how curling of the leaves helps to reduce water loss from marram grass.

______________________________________________

______________________________________________

______________________________________________ **2**

(ii) The rate of photosynthesis in marram grass leaves increases when the soil is moist.

Explain how the action of the hinge cells contributes to this increase.

______________________________________________

______________________________________________

______________________________________________ **2**

(iii) Give the term used to describe a plant which has adaptations to reduce water loss.

______________________________________________ **1**

*Marks*

**8. (continued)**

(*b*) The table below shows features found in plants adapted to grow in water.

Explain the advantages of each feature to the plants by completing the table.

| *Feature* | *Advantage to plant* |
|---|---|
| Large air spaces between leaf cells | |
| Flexible stems | |

**2**

DO NOT WRITE IN THIS MARGIN

*Marks*

**9.** The diagram below shows a section through a barley grain and the location of events occurring during germination.

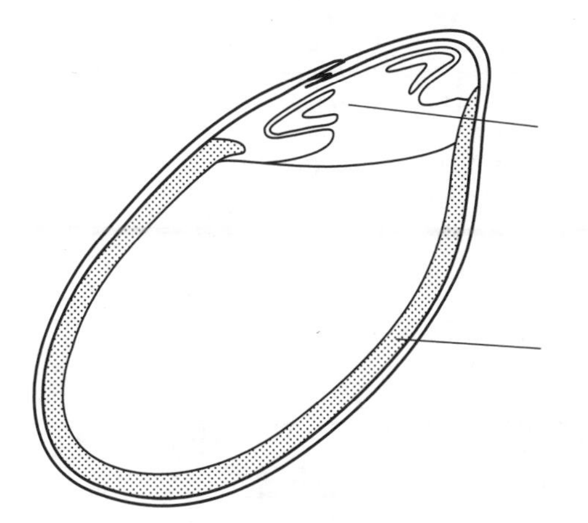


(*a*) Name growth substance P and the site of its production.

Name ______________________

Site of production ______________________ **1**

(*b*) Give the function of α-amylase and explain the importance of its action in the germination and early growth of barley grains.

Function ______________________ **1**

Importance ______________________

______________________

______________________ **1**

**[Turn over**

DO NOT WRITE IN THIS MARGIN

*Marks*

**10.** Rufous hummingbirds migrate thousands of kilometres each year between their summer breeding areas in Canada and their wintering areas in Mexico.

They feed on nectar throughout the year and are strongly territorial even on migration.

The birds save energy at night by entering a temporary state known as torpor in which body temperature and respiration rate are greatly reduced.

The **chart** below shows the average body mass of the hummingbirds and the average number of hours per night spent in torpor throughout the year.

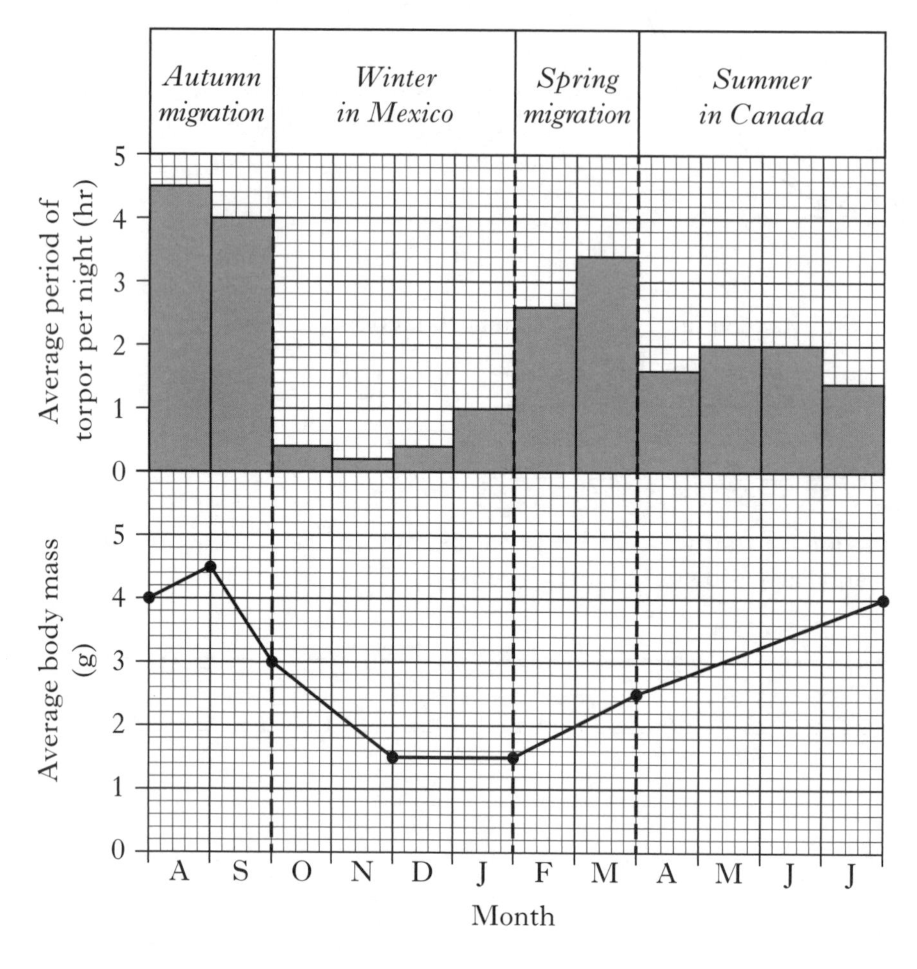


(*a*) (i) **Use values from the chart** to describe the changes in average body mass of the hummingbirds from the beginning of August until the end of January.

______________________________________________

______________________________________________

______________________________________________

______________________________________________ **2**

(ii) Calculate the percentage increase in average body mass during the summer in Canada.

*Space for calculation*

______________ % **1**

**10. (a) (continued)** *Marks*

(iii) Suggest **one** reason for the changes in average body mass of the birds during the summer in Canada.

______________________________

______________________________ **1**

(b) (i) Explain why the increased time spent in torpor during migration is an advantage to the birds.

______________________________

______________________________ **1**

(ii) Suggest a reason why the periods of torpor are longer in the summer than the winter.

______________________________

______________________________ **1**

(iii) Calculate the average period of torpor per month throughout the winter in Mexico.

*Space for calculation*

__________ hours per night **1**

(c) The **table** below shows how the average oxygen consumption of the birds at rest is affected by their body temperature.

| *Body temperature* | *Average oxygen consumption at rest* ($cm^3$ per gram of body mass per hour) |
|---|---|
| Normal | 15·0 |
| Lowered during torpor | 2·0 |

Using information from the **chart** and the **table**, calculate the average volume of oxygen consumed per hour by a hummingbird at the end of September at normal body temperature.

*Space for calculation*

__________ $cm^3$ **1**

(d) Give the advantage to rufous hummingbirds of territorial behaviour.

______________________________

______________________________ **1**

DO NOT WRITE IN THIS MARGIN

*Marks*

**11.** At the start of an investigation, the blood glucose and insulin concentrations of a healthy adult human were measured and found to be normal. The individual then immediately drank a glucose drink and his blood glucose and insulin levels were re-measured at intervals over a period of 5 hours without further food or drink intake.

The results are shown in the table below.

| *Time after glucose drink was taken* (hours) | *Glucose concentration* (mg per 100 $cm^3$) | *Insulin concentration* (units) |
|---|---|---|
| 0 (start) | 80 | 50 |
| 0·5 | 90 | 550 |
| 1 | 120 | 500 |
| 2 | 100 | 400 |
| 3 | 80 | 100 |
| 4 | 80 | 50 |
| 5 | 70 | 45 |

(*a*) Calculate the simplest whole number ratio of blood glucose concentration at the start to the maximum level recorded.

*Space for calculation*

__________ at start : __________ at maximum level **1**

(*b*) Calculate how long it took for blood insulin concentration to return to the start level from its maximum concentration.

*Space for calculation*

__________ hours **1**

(*c*) Give **two** reasons to account for the decrease in blood glucose concentration between 1 and 3 hours.

1 ____________________

____________________

2 ____________________

____________________ **2**

(*d*) Predict how the individual's blood glucagon concentration will change after 5 hours assuming no further intake of food or drink. Explain the importance of this.

Prediction ____________________

Explanation ____________________

____________________ **2**

DO NOT WRITE IN THIS MARGIN

*Marks*

**12.** (*a*) The graph below shows the changes in body mass and mass of growth hormone (GH) in the blood of a human from birth to age 24 years.

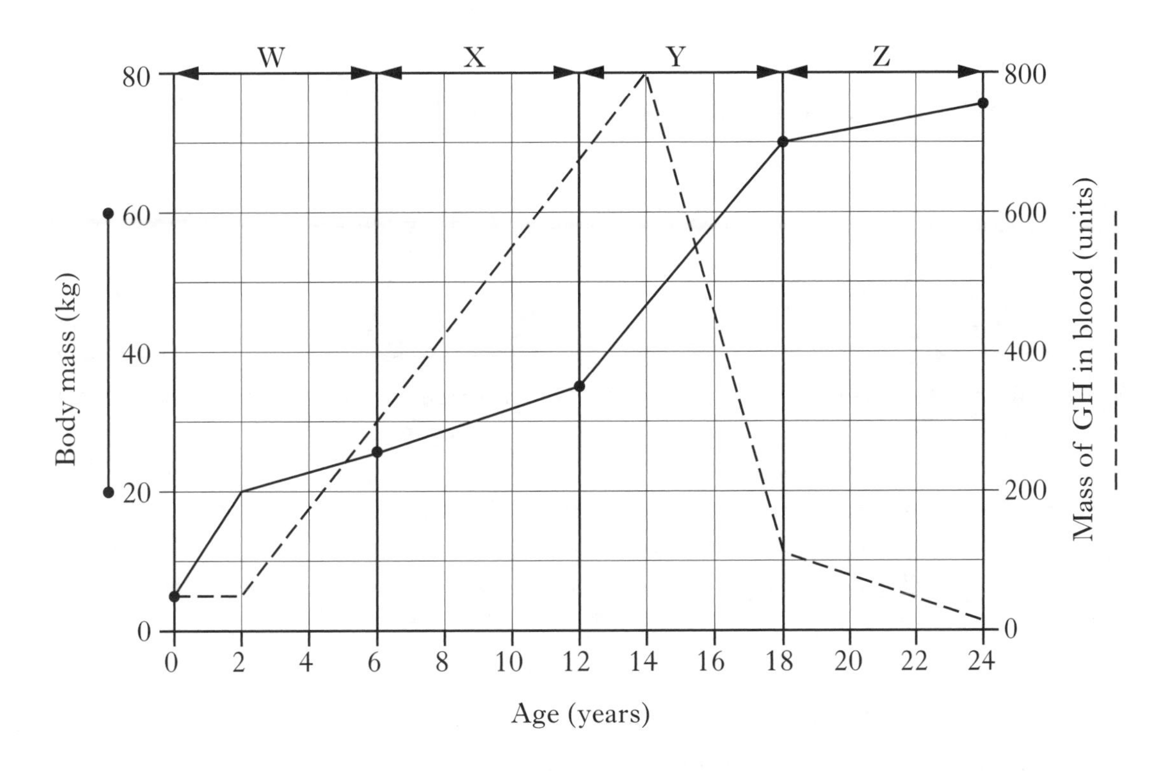


(i) Tick (✓) the box to show the age range during which the most rapid increase in body mass occurred.

0–2 years ☐ 2–12 years ☐ 12–18 years ☐ 18–24 years ☐ **1**

(ii) An increase in growth hormone (GH) causes an increase in mass of muscle and bone tissues.

Tick (✓) the box to show the region of the graph which **best** supports this statement.

W ☐ X ☐ Y ☐ Z ☐ **1**

(iii) Factors other than growth hormone (GH) are involved in increases in body mass in humans.

Describe how the graph supports this statement.

________________________________________

________________________________________ **1**

(*b*) Name the site of the production of growth hormone (GH) in humans.

________________________________________ **1**

DO NOT WRITE IN THIS MARGIN

*Marks*

**13.** In an experiment to investigate the Jacob-Monod hypothesis of lactose metabolism in *E. coli*, flasks were set up as shown in the diagram below.

Flask 1

Cotton wool plug

*E. coli* + nutrient solution with lactose

Flask 2

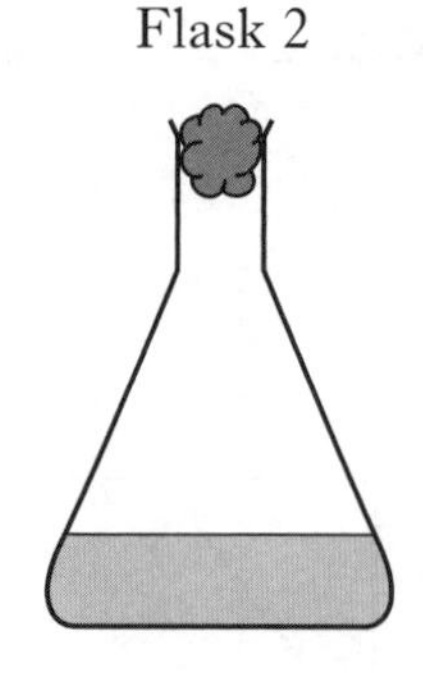

*E. coli* + nutrient solution without lactose

*E. coli* breaks down lactose using the enzyme β–galactosidase as shown.

$$\text{Lactose} \xrightarrow{\beta\text{–galactosidase}} \text{glucose + galactose}$$

(*a*) (i) β–galactosidase was produced in **Flask 1**.

Describe events which led to the production of this enzyme. **2**

(ii) β–galactosidase was **not** produced in **Flask 2**.

Explain the advantage to *E. coli* of **not** producing the enzyme in this case. **1**

(*b*) ONPG is a colourless substance which is converted to a yellow product by β–galactosidase.

Use this information to describe how ONPG could be used to show that *E. coli* only produced β–galactosidase in the presence of lactose. **1**

DO NOT WRITE IN THIS MARGIN

*Marks*

**14.** In an investigation, young plant shoots were exposed to 48 hours of light from above or from one side only.

Their growth responses are shown in the diagrams below.

| | Shoot A | Shoot B |
|---|---|---|
| | Light from above 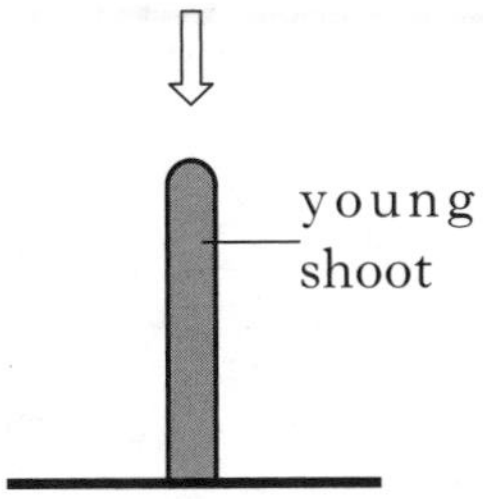 | Light from one side only 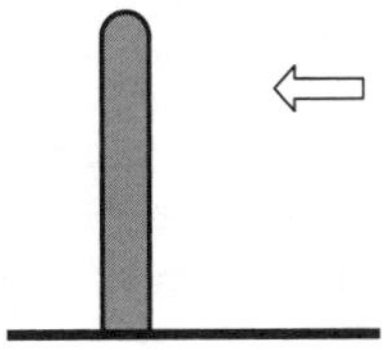 |
| At start | young shoot | |
| After 48 hours | 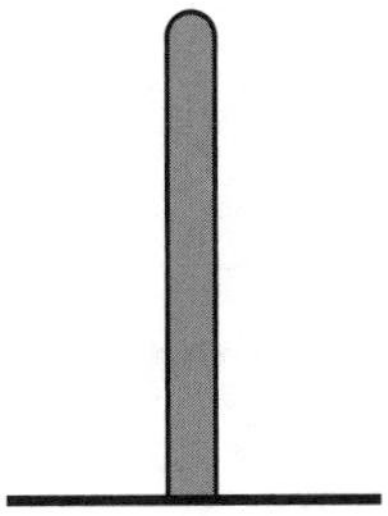 | 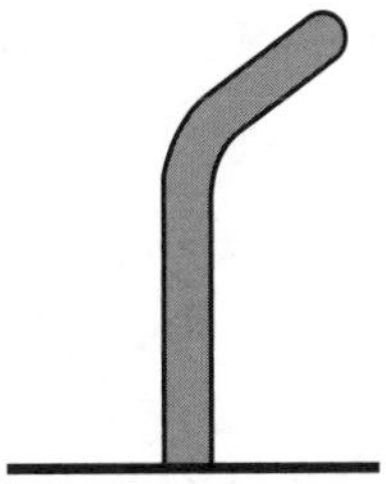 |

(*a*) Name the response shown by the shoots and explain the advantage of this response to the plants.

Name ________________________________________ **1**

Advantage ________________________________________

________________________________________ **1**

(*b*) (i) Light from one side causes elongating cells to receive an uneven distribution of a growth substance produced by the shoot tip.

Use this information to explain the growth response of **shoot B** after 48 hours.

________________________________________

________________________________________

________________________________________ **2**

(ii) Name the growth substance produced by the shoot tip which is involved in this growth response to light.

________________________________________ **1**

**[Turn over for SECTION C on *Page thirty-two***

*Marks*

## SECTION C

**Both questions in this section should be attempted.**

Note that each question contains a choice.

**Questions 1 and 2 should be attempted on the blank pages which follow.**

**Supplementary sheets, if required, may be obtained from the Invigilator.**

**All answers must be written clearly and legibly in ink.**

**Labelled diagrams may be used where appropriate.**

**1.** Answer **either** A **or** B.

**A.** Write notes on photosynthesis under the following headings:

(i) role of light and photosynthetic pigments; **4**

(ii) light dependent stage. **6**

**(10)**

**OR**

**B.** Write notes on proteins under the following headings:

(i) translation of mRNA in protein synthesis; **7**

(ii) the types and functions of protein. **3**

**(10)**

**In question 2, ONE mark is available for coherence and ONE mark is available for relevance.**

**2.** Answer **either** A **or** B.

**A.** Give an account of the regulation of blood water content in mammals following a decrease in blood water concentration. **(10)**

**OR**

**B.** Give an account of the process of succession in plant communities and the reasons for monitoring wild populations. **(10)**

[*END OF QUESTION PAPER*]

DO NOT WRITE IN THIS MARGIN

**SPACE FOR ANSWERS**

DO NOT WRITE IN THIS MARGIN

**SPACE FOR ANSWERS**

DO NOT WRITE IN THIS MARGIN

**SPACE FOR ANSWERS**

**SPACE FOR ANSWERS**

DO NOT WRITE IN THIS MARGIN

**SPACE FOR ANSWERS**

DO NOT WRITE IN THIS MARGIN

SPACE FOR ANSWERS

DO NOT WRITE IN THIS MARGIN

SPACE FOR ANSWERS

**SPACE FOR ANSWERS**

ADDITIONAL GRAPH PAPER FOR QUESTION 4 (*b*)

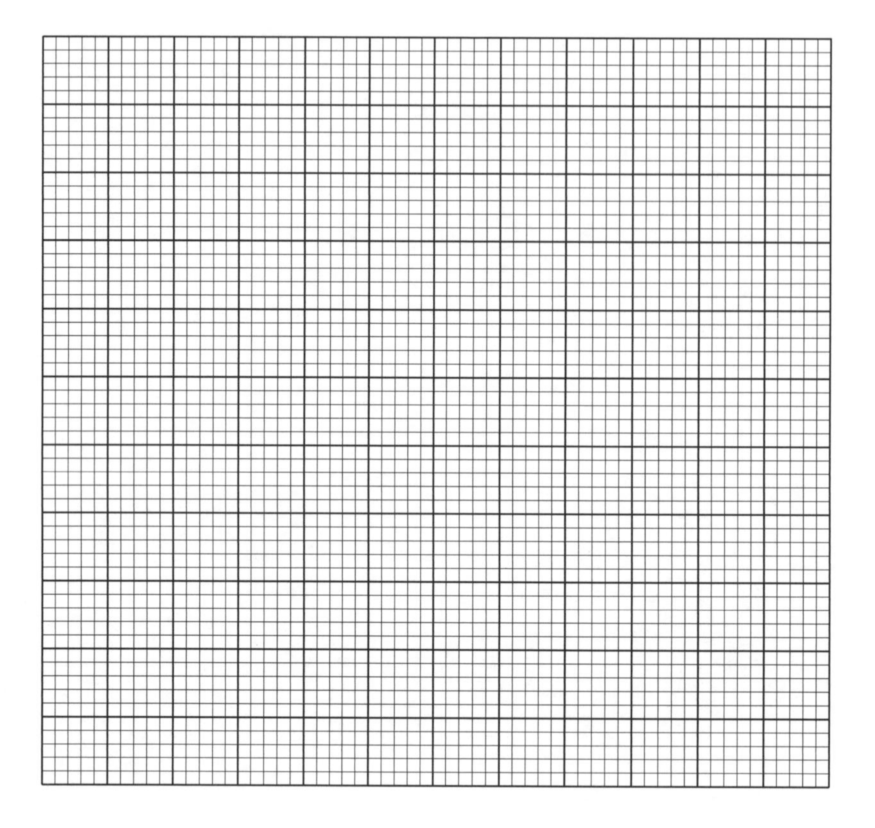

HIGHER

# 2014

[BLANK PAGE]

FOR OFFICIAL USE

| | | | | | |
|---|---|---|---|---|---|
| | | | | | |

Total for Sections B and C

# X007/12/02

NATIONAL QUALIFICATIONS 2014

FRIDAY, 16 MAY
1.00 PM – 3.30 PM

# BIOLOGY HIGHER

**Fill in these boxes and read what is printed below.**

Full name of centre

Town

Forename(s)

Surname

Date of birth
Day Month Year

Scottish candidate number

Number of seat

**SECTION A—Questions 1–30 (30 Marks)**

Instructions for completion of Section A are given on *Page two.*

For this section of the examination you must use an **HB pencil**.

**SECTIONS B AND C (100 Marks)**

1 (a) All questions should be attempted.

(b) It should be noted that in **Section C** questions 1 and 2 each contain a choice.

2 The questions may be answered in any order but all answers are to be written in the spaces provided in this answer book, **and must be written clearly and legibly in ink**.

3 Additional space for answers will be found at the end of the book. If further space is required, supplementary sheets may be obtained from the Invigilator and should be inserted inside the **front** cover of this book.

4 The numbers of questions must be clearly inserted with any answers written in the additional space.

5 Rough work, if any should be necessary, should be written in this book and then scored through when the fair copy has been written. If further space is required, a supplementary sheet for rough work may be obtained from the Invigilator.

6 Before leaving the examination room you must give this book to the Invigilator. If you do not, you may lose all the marks for this paper.

## Read carefully

1 Check that the answer sheet provided is for **Biology Higher (Section A)**.

2 For this section of the examination you must use an **HB pencil**, and where necessary, an eraser.

3 Check that the answer sheet you have been given has **your name**, **date of birth**, **SCN** (Scottish Candidate Number) and **Centre Name** printed on it.

Do not change any of these details.

4 If any of this information is wrong, tell the Invigilator immediately.

5 If this information is correct, **print** your name and seat number in the boxes provided.

6 The answer to each question is **either** A, B, C or D. Decide what your answer is, then, using your pencil, put a horizontal line in the space provided (see sample question below).

7 There is **only one correct** answer to each question.

8 Any rough working should be done on the question paper or the rough working sheet, **not** on your answer sheet.

9 At the end of the examination, put the **answer sheet for Section A inside the front cover of this answer book**.

## Sample Question

The apparatus used to determine the energy stored in a foodstuff is a

A calorimeter

B respirometer

C klinostat

D gas burette.

The correct answer is **A**—calorimeter. The answer **A** has been clearly marked in **pencil** with a horizontal line (see below).

A B C D

## Changing an answer

If you decide to change your answer, carefully erase your first answer and using your pencil fill in the answer you want. The answer below has been changed to **D**.

A B C D

## SECTION A

**All questions in this section should be attempted.**

**Answers should be given on the separate answer sheet provided.**

**1.** The action spectrum of photosynthesis shows the ability of green plants to

A use light for photolysis

B absorb all wavelengths of light in photosynthesis

C absorb different wavelengths of light in photosynthesis

D use light of different wavelengths for photosynthesis.

**2.** The graph below shows the uptake of potassium ions by carrot tissue at different temperatures.

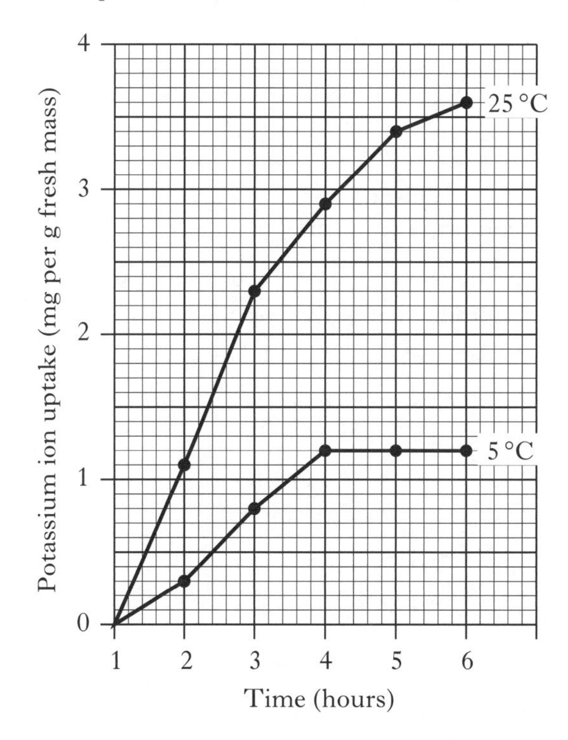


At 6 hours, how many times greater is the uptake of potassium ions at 25 °C compared with 5 °C?

A 1·2

B 2·0

C 2·4

D 3·0

**3.** An investigation was carried out to compare photosynthesis in green light by oak and nettle leaves. Five leaf discs were cut from each plant and placed in syringes containing a solution to provide carbon dioxide.

The diagram below shows the positions of the leaf discs after one hour.

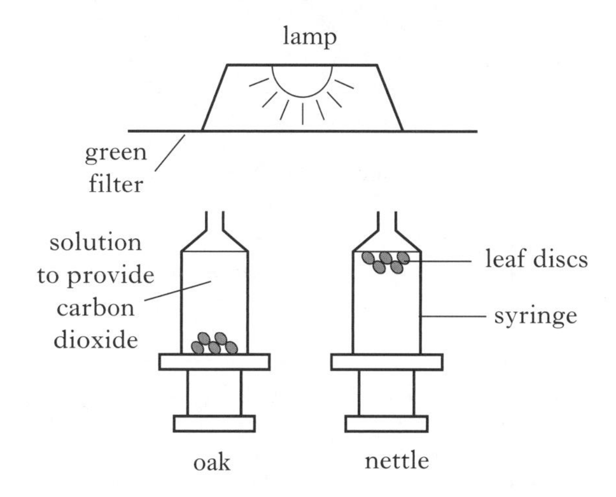


How could the investigation be improved to allow valid conclusions to be drawn?

A Carry out the experiment in a darkened room.

B Use different species of plant.

C Use more leaf discs.

D Repeat the experiment.

[Turn over

**4.** The diagram below represents a stage in protein synthesis in a cell.

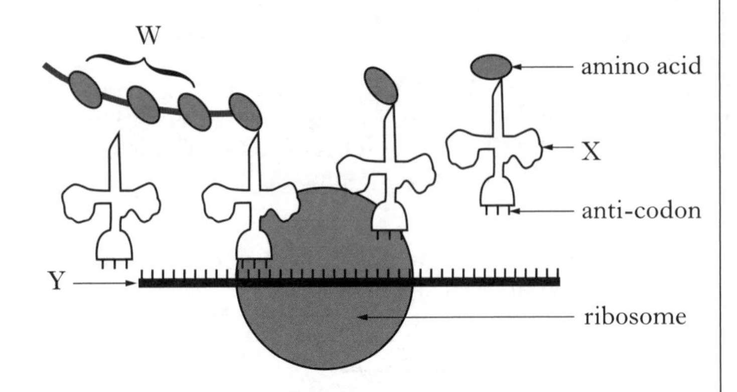


Which line in the table below identifies correctly molecules W, X and Y?

| | *Molecules* | | |
|---|---|---|---|
| | W | X | Y |
| A | tRNA | mRNA | polypeptide |
| B | polypeptide | tRNA | mRNA |
| C | mRNA | tRNA | polypeptide |
| D | polypeptide | mRNA | tRNA |

**5.** Which line in the table below shows correctly where ATP is synthesised in a yeast cell respiring anaerobically?

| | *Site of ATP synthesis* | |
|---|---|---|
| | *Mitochondrion* | *Cytoplasm* |
| A | Yes | No |
| B | No | Yes |
| C | No | No |
| D | Yes | Yes |

**6.** Four stages in the process of phagocytosis are shown below.

1 Lysosomes fuse with vacuole

2 Bacterium becomes enclosed in vacuole

3 Bacterium is digested by enzymes

4 Breakdown products pass into cytoplasm

In which sequence do these stages occur?

A 1, 2, 3 ,4

B 2, 1, 3, 4

C 2, 3, 1, 4

D 1, 2, 4, 3

**7.** Which substances must be provided by host cells for the synthesis of viruses?

A Proteins and nucleotides

B Amino acids and DNA

C Proteins and DNA

D Amino acids and nucleotides

**8.** The following events occur during the replication of a virus.

1 Alteration of host cell metabolism

2 Production of viral protein coats

3 Replication of viral DNA

In which sequence do these events occur?

A 1 → 3 → 2

B 1 → 2 → 3

C 2 → 1 → 3

D 3 → 1 → 2

**9.** Which of the following statements about antibodies is **not** correct?

A They are specific to antigens.

B They are globular proteins.

C They are produced by phagocytes.

D They are involved in tissue rejection.

10. The graph below shows the concentration of antibodies in the blood of a patient after repeated exposures to a foreign antigen.

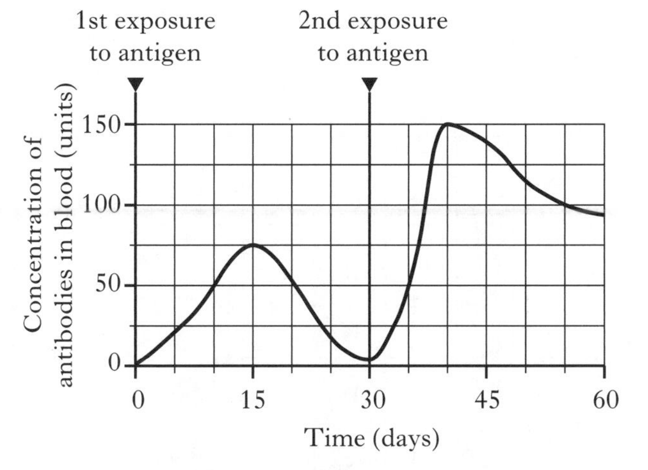


What is the percentage increase in the concentration of antibodies in the patient's blood 10 days after the second exposure compared to 10 days after the first exposure?

A 33%

B 67%

C 100%

D 200%

11. The chromosomes of a gamete mother cell are shown in the diagram below.

How many chromosomes would be present in each gamete produced?

A 2

B 4

C 8

D 16

12. Which of the following are produced by meiosis?

A Haploid cells of identical genetic composition

B Diploid cells of different genetic composition

C Diploid cells of identical genetic composition

D Haploid cells of different genetic composition

13. In a species of pea plant, dwarfness and white petal colour are caused by recessive alleles found on separate chromosomes. The corresponding dominant alleles are tallness and coloured petals.

If a dwarf plant with white petals is crossed with a plant heterozygous for each characteristic, what proportion of the offspring would be expected to be dwarf plants with white petals?

A 1 in 3

B 1 in 4

C 3 in 16

D 9 in 16

14. The inheritance of eye colour in fruit flies is sex-linked and the allele for red eyes **R** is dominant to the allele for white eyes **r**.

The offspring from a cross were all red-eyed females and white-eyed males.

What were the genotypes of the parents?

A $X^rX^r$ $X^RY$

B $X^RX^r$ $X^RY$

C $X^RX^r$ $X^rY$

D $X^RX^R$ $X^rY$

**[Turn over**

**15.** The diagram below represents four populations W, X, Y and Z of a small mammal and the areas where they interbreed.

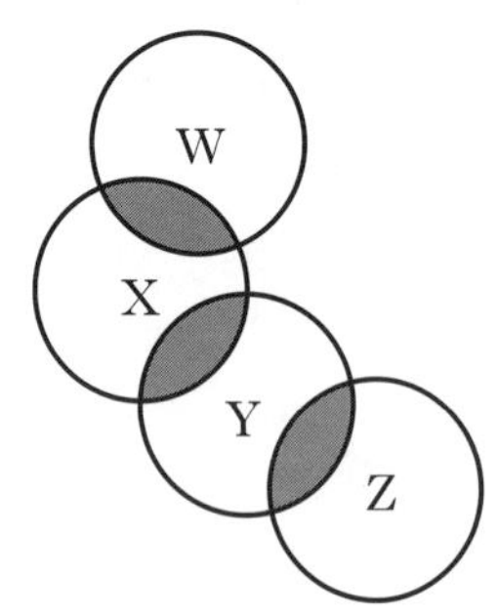


How many species are present?

A 1

B 2

C 3

D 4

**16.** Which line in the table below identifies correctly the enzyme required and structures fused together in the process of somatic fusion of plant cells?

| | *Enzyme required* | *Structures fused together* |
|---|---|---|
| A | cellulase | plasmids |
| B | cellulase | protoplasts |
| C | endonuclease | protoplasts |
| D | endonuclease | plasmids |

**17.** The larvae of gypsy moths and cotton boll worms are pests of tree leaves. An experimental plot of infested trees was sprayed with insecticide in three different years. The numbers of each larvae killed in each year is shown in the graph below.

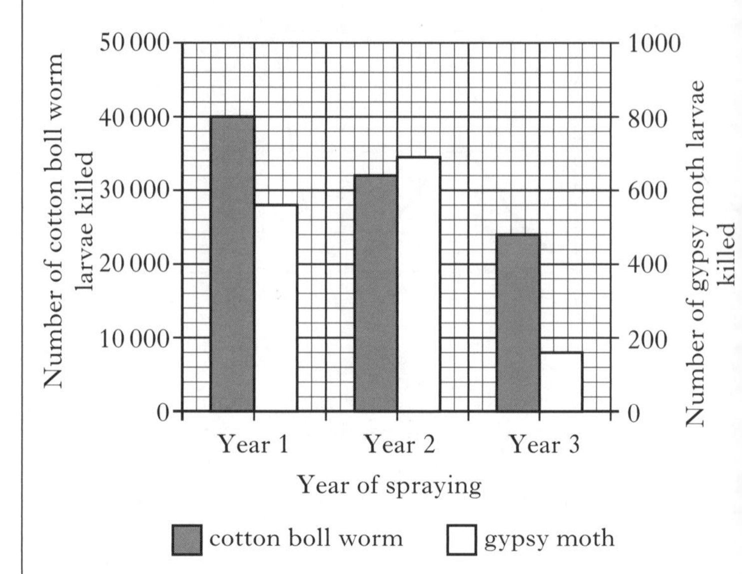


Which of the following conclusions can be drawn?

A More gypsy moth larvae were killed than cotton boll worm larvae in year 2.

B The larvae became more resistant to the insecticide each year.

C The number of gypsy moth larvae killed was always less than cotton boll worm larvae killed.

D The percentage of cotton boll worm larvae surviving decreased each year.

**18.** Huntington's disease is caused by a single dominant allele of a gene which is not sex-linked.

A woman's father is heterozygous for this condition and her mother is unaffected.

What are the chances of this woman inheriting Huntington's disease?

A 75%

B 67%

C 50%

D 25%

**19.** Leaves of the same size were cut from the same plant and treated as shown in **Diagram 1**.

**Diagram 2** shows how the apparatus appeared after being left for one hour in a warm dry environment.

**Diagram 1**
(Mass of leaves balanced)

balance

Leaf **T**

Leaf **S**

Lower surface coated with petroleum jelly to block stomata

Upper surface coated with petroleum jelly to block stomata

**Diagram 2**

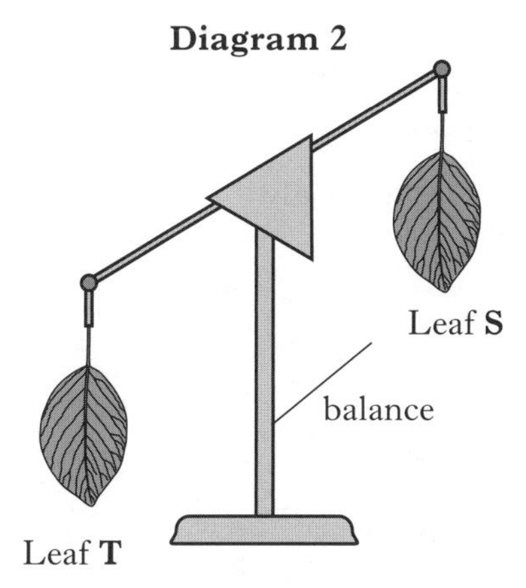


Which of the following statements relating to the results of this experiment is correct?

A Leaf **T** has more stomata than leaf **S**.

B Leaf **S** has more stomata than leaf **T**.

C The leaves have more stomata in their lower epidermis than in their upper epidermis.

D The leaves have more stomata in their upper epidermis than in their lower epidermis.

**20.** The list below shows statements about some animal species.

1 Common quail rest together in groups.
2 European hedgehogs have spines on their skin.
3 Grey wolves hunt in packs.
4 Kittiwakes nest in colonies.

Which statements describe social mechanisms for defence?

A 1 and 4 only

B 1, 3 and 4 only

C 2 and 3 only

D 2 and 4 only

**21.** The graph below shows the pattern of growth of an organism over a period of 4 months.

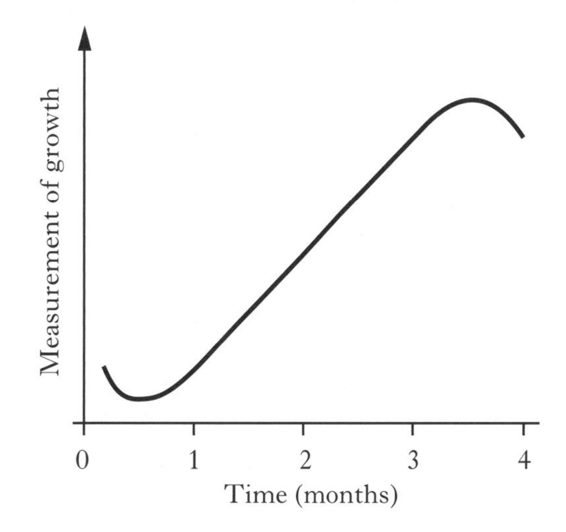


The graph shows changes in the

A mass of an insect

B dry mass of an annual plant

C length of an insect

D length of an annual plant.

**[Turn over**

**22.** Which line in the table below provides correct information on control of body temperature in mammals?

| | *Monitoring centre* | *Form of communication* | *Effector organ* |
|---|---|---|---|
| A | skin | hormonal | liver |
| B | skin | nervous | brain |
| C | hypothalamus | hormonal | liver |
| D | hypothalamus | nervous | skin |

**23.** The diagram below shows the relationship between the pituitary gland and some processes of growth and development in humans.

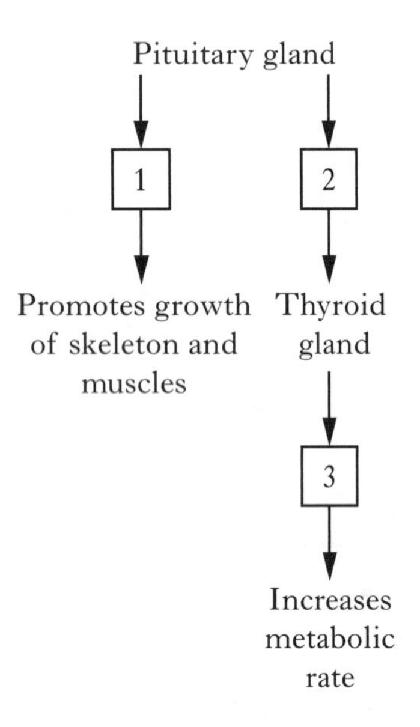


Which line in the table below identifies correctly the hormones 1 to 3?

| | 1 | 2 | 3 |
|---|---|---|---|
| A | Thyroxine | Growth hormone | TSH |
| B | TSH | Growth hormone | Thyroxine |
| C | Growth hormone | Thyroxine | TSH |
| D | Growth hormone | TSH | Thyroxine |

**24.** In a germinating barley grain, gibberellic acid (GA) stimulates the production of

A α-amylase by the aleurone layer

B α-amylase by the endosperm

C maltose by the aleurone layer

D maltose by the endosperm.

**25.** The list below shows processes which affect plants.

1 Leaf abscission

2 Fruit formation

3 Photoperiodism

4 Apical dominance

Which processes involve indole acetic acid (IAA)?

A 1 and 3 only

B 2 and 3 only

C 1, 2 and 4 only

D 3 and 4 only

**26.** In humans, vitamin D plays an essential role in the absorption of

A amino acids

B lipids

C iron

D calcium.

**27.** The flow chart below shows some events in the control of the concentration of glucose in the blood.

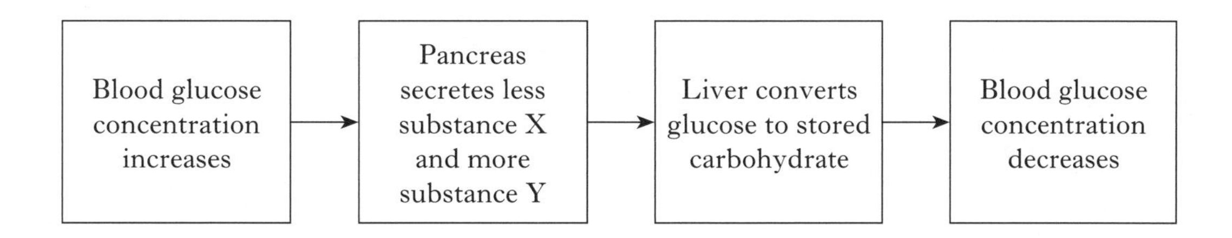


Which line in the table below identifies correctly substances X and Y?

| | *Substance X* | *Substance Y* |
|---|---|---|
| A | insulin | glucagon |
| B | insulin | glycogen |
| C | glucagon | insulin |
| D | glycogen | insulin |

**28.** The graph below shows how the rate of oxygen uptake of a tropical fish species in an aquarium varies with water temperature.

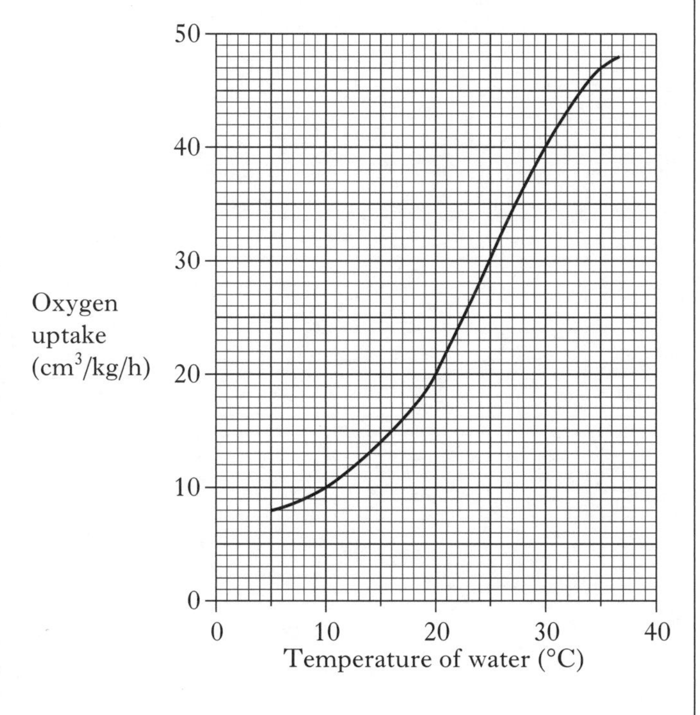


The volume of oxygen uptake for a 0·25 kg fish over a 5 hour period of time at 20 °C is

A 5 $cm^3$

B 20 $cm^3$

C 25 $cm^3$

D 100 $cm^3$.

**29.** The graph below shows the effect of population density on the number of eggs laid per day by female flour beetles.

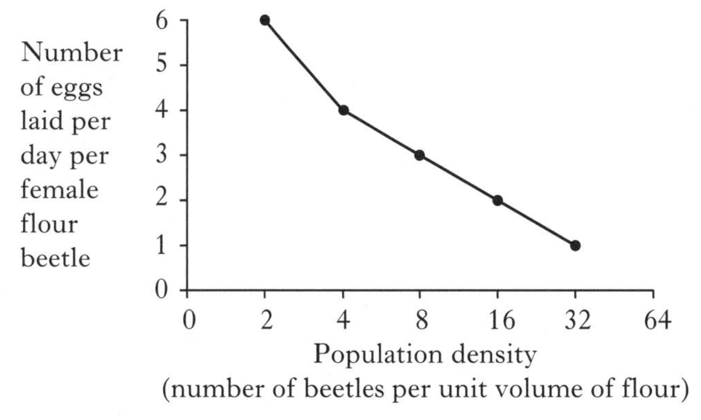


From the graph, the following conclusions were suggested.

1 Females lay fewer eggs per day as the population density increases.

2 Females lay more eggs per day at low population densities.

3 As the population density decreases, females lay fewer eggs per day.

Which of the conclusions are correct?

A 1 and 2 only

B 1 and 3 only

C 2 and 3 only

D 1, 2 and 3

**30.** Which of the following factors that can limit population size is density dependent?

A Decrease in food supply

B Cold winter

C Volcanic eruption

D Pollution

**Candidates are reminded that the answer sheet MUST be returned INSIDE the front cover of this answer book.**

**[Turn over for Section B on *Page twelve***

DO NOT WRITE IN THIS MARGIN

## SECTION B

**All questions in this section should be attempted.**

**All answers must be written clearly and legibly in ink.**

*Marks*

**1.** The diagram below shows an outline of three stages of aerobic respiration in muscle cells.

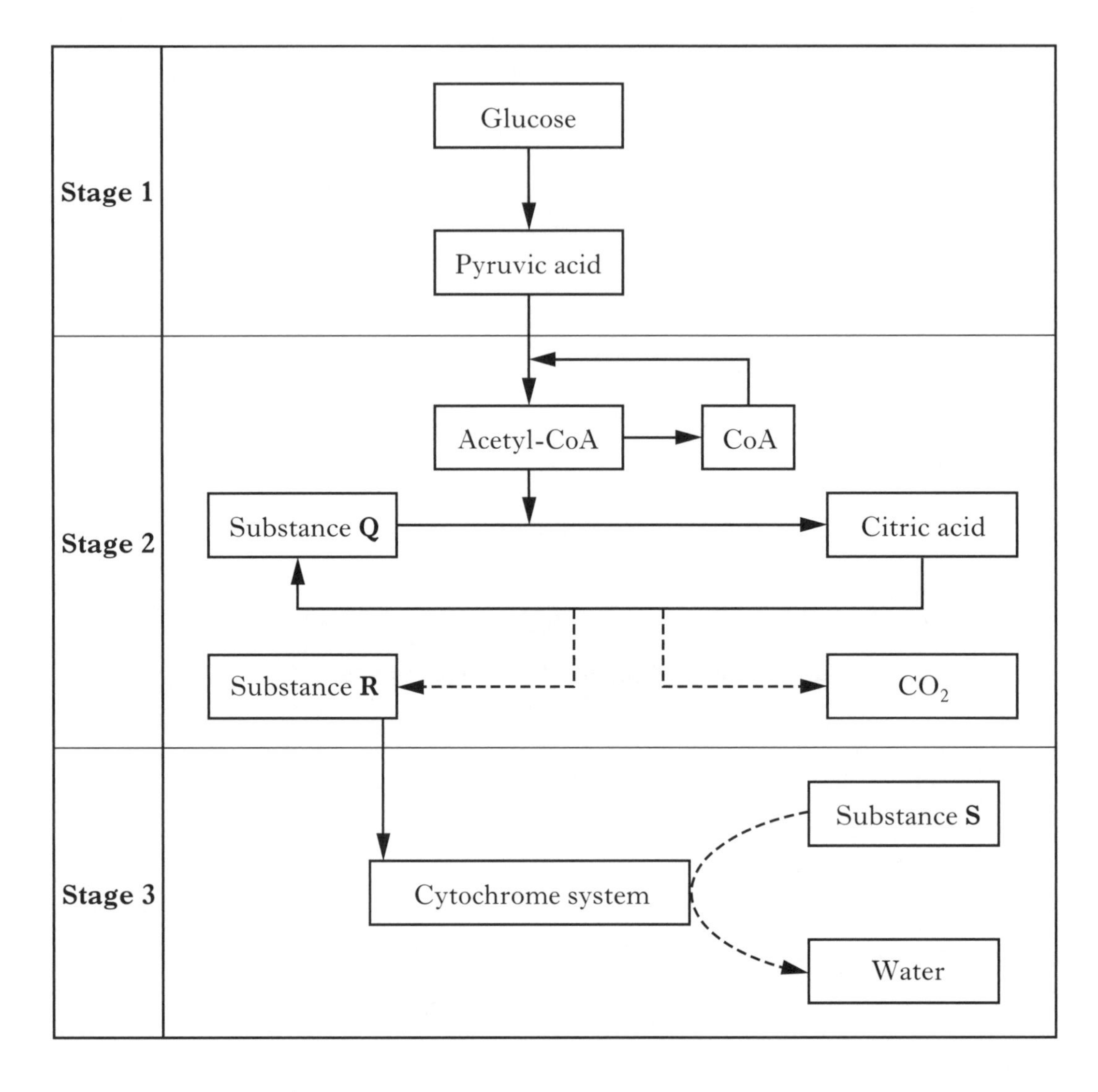


(*a*) Name **Stage 1**.

______________________________ **1**

(*b*) Complete the table below to show the number of carbon atoms in one molecule of each substance.

| *Substance* | *Number of carbon atoms in one molecule* |
|---|---|
| Pyruvic acid | |
| Substance **Q** | |
| Citric acid | |

**2**

DO NOT WRITE IN THIS MARGIN

*Marks*

**1. (continued)**

(*c*) Name Substance **R** in **Stage 2** and the carrier which transfers it to the cytochrome system.

Substance **R** ______________________________

Carrier ______________________________ **1**

(*d*) Name Substance **S** in **Stage 3** and its role in aerobic respiration.

(i) Substance **S** ______________________________ **1**

(ii) Role in aerobic respiration ______________________________

______________________________ **1**

(*e*) The diagrams below show the structure of a mitochondrion from a skin cell and one from a muscle cell.

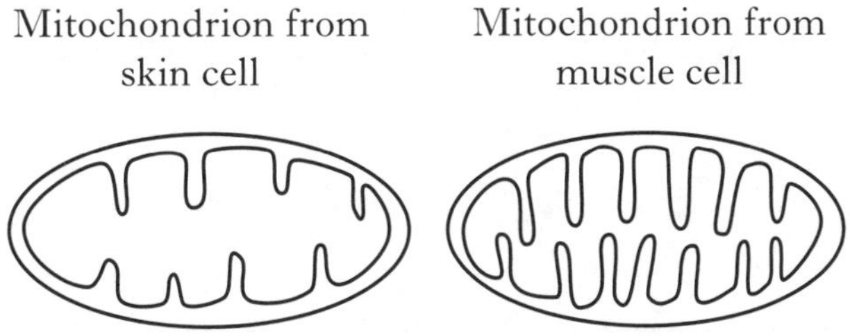

Mitochondrion from skin cell

Mitochondrion from muscle cell

Describe the difference in structure between the two mitochondria and explain how this is related to the function of the muscle cell.

Difference ______________________________

______________________________

Explanation ______________________________

______________________________ **2**

**[Turn over**

DO NOT WRITE IN THIS MARGIN

*Marks*

**2.** Nettles are shade plants which often grow below trees.

Their leaves contain photosynthetic pigments X and Y. The table below shows the percentage of light of different wavelengths absorbed by these pigments.

| *Wavelength of light* (nm) | *Colour of light* | *Light absorbed* (%) | |
|---|---|---|---|
| | | *Pigment X* | *Pigment Y* |
| 400 | violet | 40 | 20 |
| 440 | blue | 60 | 30 |
| 550 | green | 5 | 60 |
| 680 | red | 50 | 5 |

(*a*) Apart from being absorbed, state what else can happen to light striking the leaves of plants.

______________________________ **1**

(*b*) Identify which of the pigments, X or Y, in the table is chlorophyll.

Justify your choice.

Pigment ____________

Justification ______________________________

______________________________ **1**

(*c*) (i) Describe the relationship between the wavelength of light and the percentage of light absorbed by pigment Y.

______________________________

______________________________

______________________________ **2**

(ii) Describe how the presence of pigment Y in their leaves would benefit nettle plants growing below trees.

______________________________

______________________________

______________________________ **1**

DO NOT WRITE IN THIS MARGIN

*Marks*

**3.** (*a*) Decide if each of the statements relating to DNA in the table below is **true** or **false** and tick (✓) the appropriate box.

If you decide that the statement is **false**, write the correct term in the correction box to replace the term underlined in the statement.

| *Statement* | *True* | *False* | *Correction* |
|---|---|---|---|
| The region of a DNA molecule which codes for a protein is called a gene. | | | |
| Pores in the nuclear membrane allow DNA to carry the code for a protein out of the nucleus. | | | |
| A DNA molecule has many codons each made up of deoxyribose, phosphate and a base. | | | |

2

(*b*) A section of a DNA molecule containing a total of 1600 bases has 184 adenine and 216 thymine bases on one strand. The complementary strand contains 268 cytosine bases.

(i) Calculate the number of adenine bases in the **whole section** of the DNA molecule.

*Space for calculation*

________________ 1

(ii) Calculate the number of guanine bases in the **complementary** strand.

*Space for calculation*

________________ 1

(*c*) Name the structure in cells which transports protein from the ribosomes to the Golgi apparatus.

________________________________ 1

DO NOT WRITE IN THIS MARGIN

*Marks*

**3. (continued)**

(*d*) The list below shows different proteins.

1 cellulase

2 collagen

3 insulin

4 endonuclease

Use numbers from the list to complete the table below.

| *Class of protein* | |
|---|---|
| *globular* | *fibrous* |
| | |

2

DO NOT WRITE IN THIS MARGIN

Marks

**4.** A culture of yeast was grown in 5 litres of glucose solution.

Glucose and alcohol concentrations in the culture were measured every 5 hours for 25 hours.

The results are shown in the graph below.

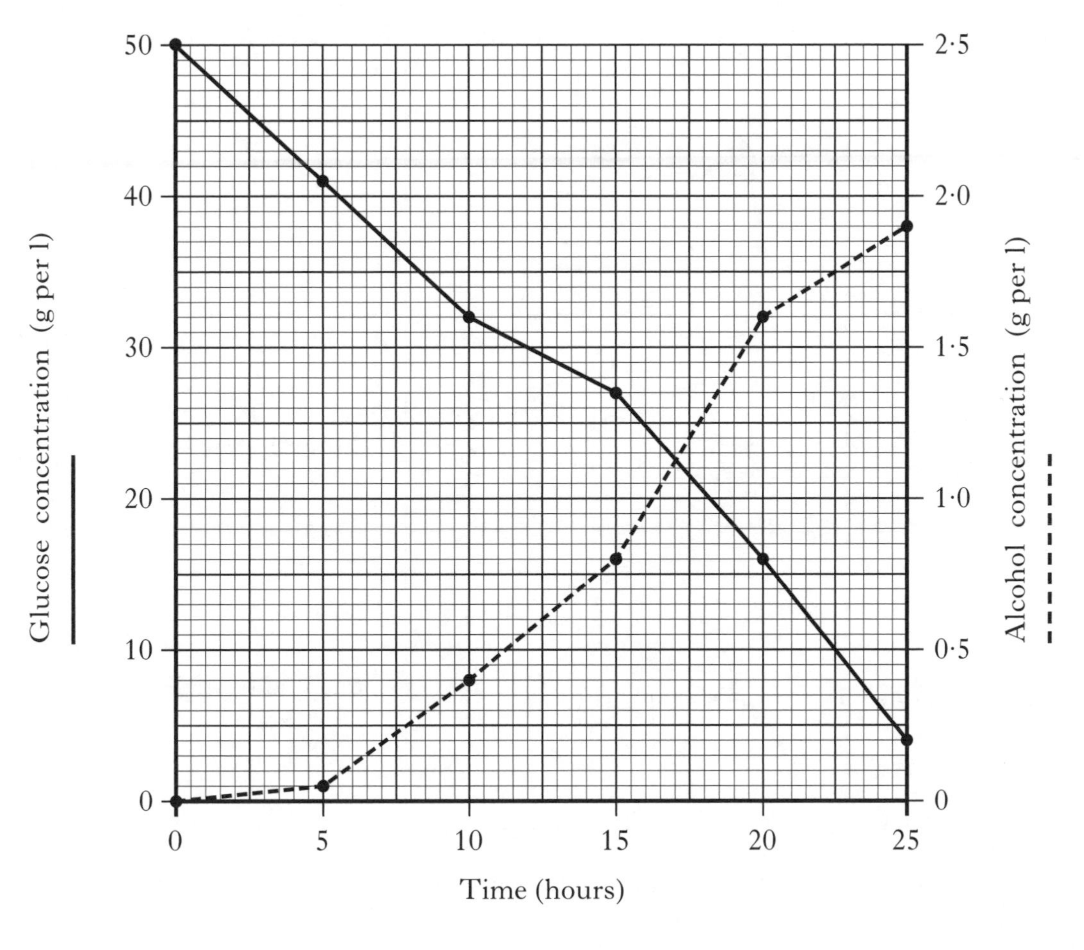


(*a*) State the alcohol concentration when the glucose concentration was 10 g per litre.

_______________ g per litre **1**

(*b*) Tick (✓) the box to identify the time period during which the rate of alcohol production was the greatest.

| 0 – 10 hours | 5 – 15 hours | 10 – 20 hours | 15 – 25 hours |
|---|---|---|---|
| ☐ | ☐ | ☐ | ☐ |

**1**

(*c*) Identify the time at which the glucose concentration reached 50% of its starting concentration.

_______________ hours **1**

(*d*) Calculate how many grams of glucose remained in the solution at the end of the investigation.

*Space for calculation*

_______________ g **1**

DO NOT WRITE IN THIS MARGIN

*Marks*

**5.** In North America, three species of *Penstemon* plants have evolved from a single species. The plants are pollinated by animals that feed on nectar found in nectaries inside their flower tubes, close to the base.

| *Species* | *Scale drawing of flower tube* | *Specialised pollinator* |
|---|---|---|
| *Penstemon grinnellii* | | large bees |
| *Penstemon centranthifolius* | | hummingbirds |
| *Penstemon spectabilis* | | wasps and small bees |

(*a*) Use the information above to explain how the evolution of the *Penstemon* species illustrates adaptive radiation. **2**

(*b*) Several species of hummingbird compete for *Penstemon centranthifolius* nectar.

Name the type of competition involved in this example. **1**

(*c*) The various pollinators must forage economically.

Explain what is meant by this statement in terms of energy. **1**

**[Turn over for Question 6 on *Page twenty***

DO NOT WRITE IN THIS MARGIN

*Marks*

**6.** Oil from wild varieties of oilseed rape plants contains a high concentration of erucic acid which makes the oil unsuitable for human consumption.

Selective breeding programmes have produced modern varieties of oilseed rape plants with oil of low erucic acid concentration which is suitable for human consumption.

In 2003 a new selective breeding programme was started which aimed to further reduce the erucic acid concentration of the oil and to increase oil content of seeds.

The **bar chart** below shows the results of the new selective breeding programme over a 10 year period.

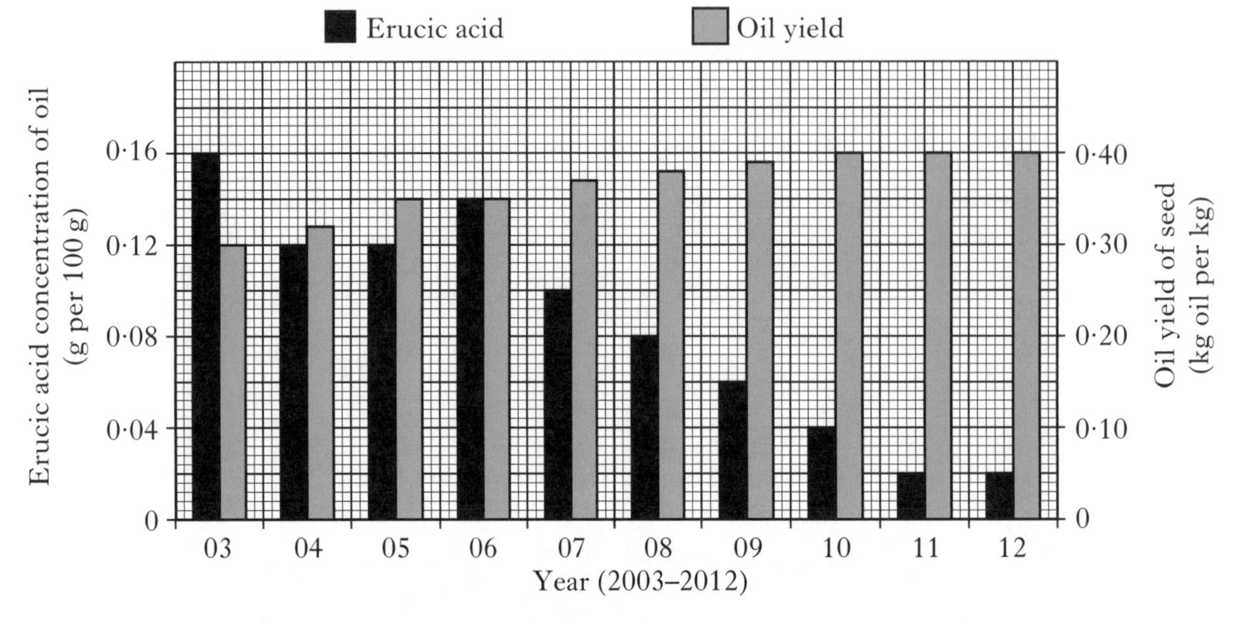


(*a*) (i) Use values from the **bar chart** to describe the changes in erucic acid concentration of the oil from 2005 until 2012.

________________________________________

________________________________________

________________________________________

________________________________________ **2**

(ii) Calculate the simplest whole number ratio of the erucic acid concentration in the 2003 harvest compared with that of the 2011 harvest.

*Space for calculation*

_______ in 2003 : _______ in 2011 **1**

(iii) Calculate the mass of seed from 2012 which would be needed to produce one kilogram of oil.

*Space for calculation*

_______ kg **1**

DO NOT WRITE IN THIS MARGIN

*Marks*

**6. (continued)**

(*b*) The bacterium *Bacillus thuringiensis* produces Bt-toxin, a substance harmful to leaf-eating insects. Some oilseed rape plants were genetically engineered so that they contained the gene for Bt-toxin.

A field trial was set up to compare seed yields in genetically engineered plants with the Bt-toxin gene and control plants without the Bt-toxin gene. Equal numbers of the two types of plant were grown under identical conditions in the presence of leaf-eating insects and their seed yield per hectare compared.

The results of the trial are shown in the **table** below.

| *Plants* | *Seed yield* (kg per hectare) |
|---|---|
| Genetically engineered (with the Bt-toxin gene) | 144 |
| Control (without the Bt-toxin gene) | 80 |

(i) Calculate the percentage increase in the seed yield per hectare from plants with the Bt-toxin gene compared with the control plants.

*Space for calculation*

______ % **1**

(ii) Explain why the genetically engineered plants produce a higher yield of seed per hectare compared with the control plants.

______

______

______ **2**

(iii) The selectively bred plants which produced the 2012 harvest were affected by leaf-eating insects.

Using information from the **table** and the **bar chart**, predict the increase in **oil yield** per hectare which could have been achieved, if these plants had been:

- genetically engineered to contain the Bt-toxin gene
- grown under identical conditions to those in the field trial.

*Space for calculation*

Increase in oil yield: ______ kg oil per hectare **1**

**[Turn over**

DO NOT WRITE IN THIS MARGIN

*Marks*

**7.** The bacterium *Escherichia coli* lives in the intestines of domestic pigs. New generations of *E. coli* can arise every twenty minutes under ideal conditions.

An investigation was carried out over a six-month period during which time pigs were regularly injected with a mixture of antibiotics A and B.

At regular intervals during this investigation, *E. coli* from the intestines of the pigs were sampled and tested for antibiotic resistance.

The graph below shows changes in the percentage of *E. coli* in the samples which were resistant to each antibiotic.

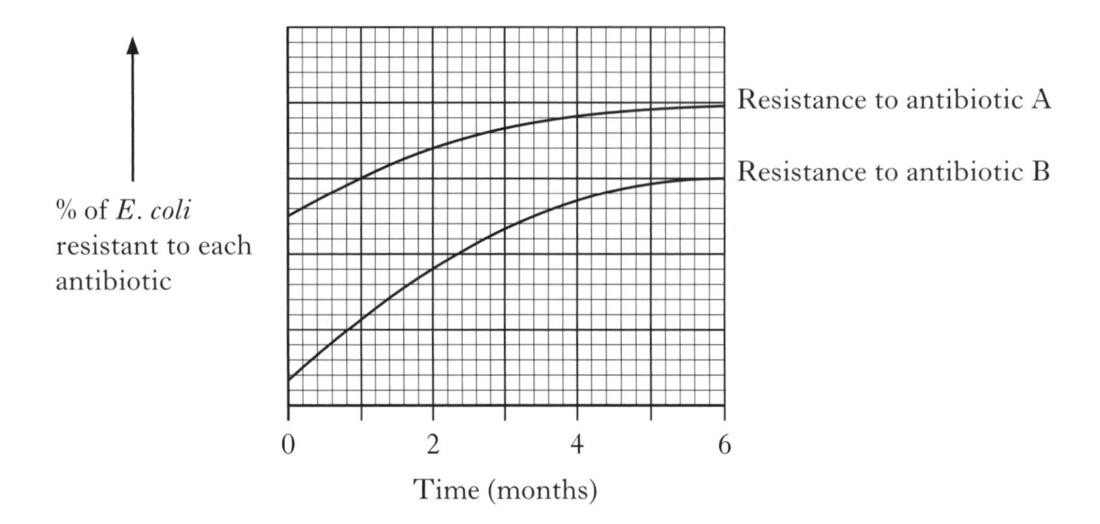


(*a*) Explain how the increase in resistance to antibiotic B has arisen by natural selection.

______________________________________________

______________________________________________

______________________________________________ **2**

(*b*) The pigs in this investigation had previously been treated with one of the antibiotics.

Identify this antibiotic and justify your answer with evidence from the graph.

Antibiotic ________

Justification ______________________________________

______________________________________________ **1**

DO NOT WRITE IN THIS MARGIN

*Marks*

**8.** The graph below shows how the intensity of grazing by rabbits affects the diversity of plant species in an area of grassland.

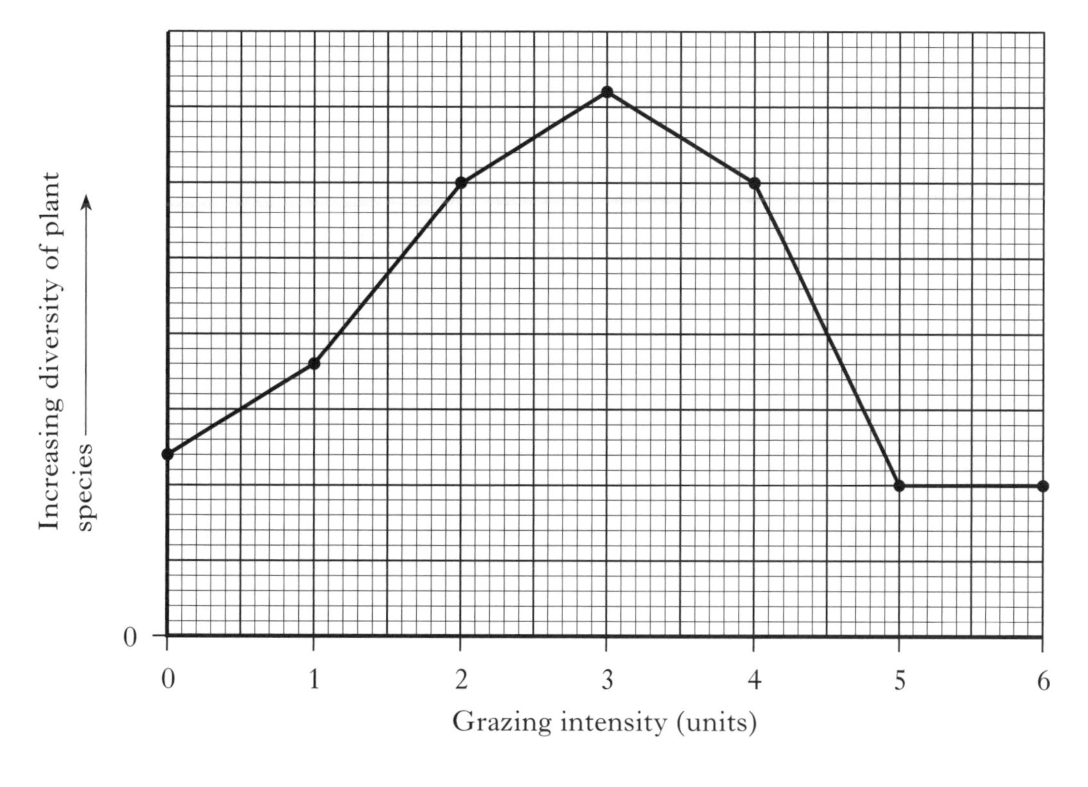


(*a*) Explain the change in diversity of plant species shown on the graph between 0 and 3 units of grazing intensity. **2**

(*b*) (i) Give evidence from the graph that indicates that the grassland contained some plant species that are tolerant of grazing. **1**

(ii) Give **one** adaptation that allows plants to tolerate grazing. **1**

**[Turn over**

DO NOT WRITE IN THIS MARGIN

*Marks*

**8. (continued)**

(*c*) Some plants have adaptations that allow them to discourage grazing.

(i) Name a structural defence mechanism that allows some species to discourage grazing.

________________________________________ **1**

(ii) Name a substance produced by the cells of some species that discourages grazing.

________________________________________ **1**

Marks

**9.** (*a*) The table below relates to adaptations for maintaining water balance in desert rats.

Complete the table by adding information to each empty box.

| *Type of adaptation* | *Description of adaptation* | *Effect of adaptation* |
|---|---|---|
| physiological | | avoids water loss by evaporation from skin |
| | nocturnal | reduces water loss by evaporation during the day |
| | | increases reabsorption of water by the kidney |

**2**

(*b*) The grid below shows the adaptations of bony fish for osmoregulation.

| | |
|---|---|
| A<br>many,<br>large glomeruli | B<br>few,<br>small glomeruli |
| C<br>low kidney<br>filtration rate | D<br>high kidney<br>filtration rate |
| E<br>chloride secretory cells<br>absorb salts | F<br>chloride secretory cells<br>secrete salts |

(i) Give **three** letters from the grid which identify adaptations of salt water bony fish.

________ , ________ and ________ **1**

(ii) Give **two** letters from the grid which identify adaptations leading to the production of high volumes of urine.

________ and ________ **1**

DO NOT WRITE IN THIS MARGIN

*Marks*

**10.** (*a*) The diagram below shows part of a section through a woody stem of a pine tree enlarged to show some of the vessels which make up the xylem tissue.

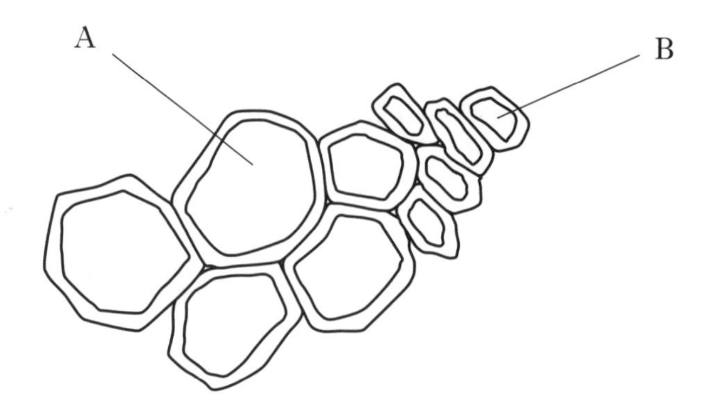


(i) Give the letter which identifies the xylem vessel formed in spring.

Give a reason for your answer.

Letter __________

Reason ____________________________________

__________________________________________ **1**

(ii) Name the meristem responsible for producing cells which differentiate into xylem vessels in a woody stem.

______________________________ **1**

(iii) Name the area in a cross section of a woody stem that represents the xylem growth occurring in one year.

______________________________ **1**

(*b*) Explain the production of differentiated tissue such as xylem in terms of gene activity.

__________________________________________

__________________________________________

__________________________________________ **1**

DO NOT WRITE IN THIS MARGIN

*Marks*

**11.** The diagram below shows information relating to the Jacob-Monod hypothesis for the control of gene action in the bacterium *Escherichia coli*.

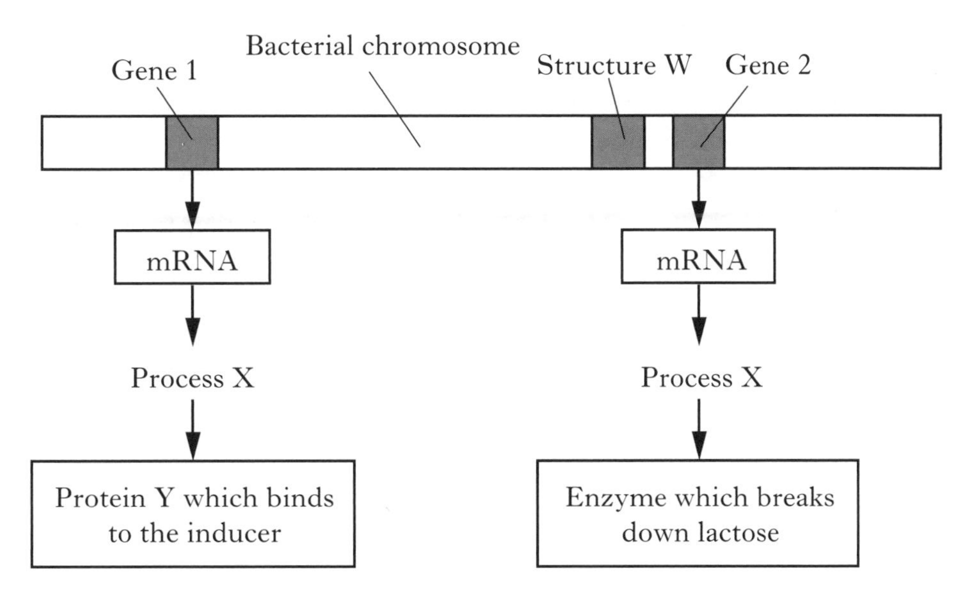


(*a*) Complete the table below.

| *Structure* | *Name* | *Function* |
|---|---|---|
| Gene 1 | | Codes for protein Y |
| Structure W | operator | |
| Gene 2 | | Codes for the enzyme which breaks down lactose |

**2**

(*b*) Name process X.

____________________________ **1**

(*c*) (i) Name the inducer.

____________________________ **1**

(ii) Explain the advantage to *E. coli* of this type of control of gene action.

____________________________

____________________________ **1**

DO NOT WRITE IN THIS MARGIN

*Marks*

**12.** An investigation was carried out to study the effects of exercise on sweat production in humans.

An exercise bike was placed in a laboratory with constant humidity and temperature.

A healthy 30-year-old male exercised on the bike for five trials of different durations as shown in the table below. The average rate of sweat production during each trial was calculated.

There was a recovery period after each trial to allow sweat production to return to normal level.

The results are shown in the table below.

| *Exercise trial* | *Duration of exercise trial* (s) | *Average rate of sweat production* (mg per $cm^2$ skin per minute) |
|---|---|---|
| 1 | 30 | 0·10 |
| 2 | 60 | 0·21 |
| 3 | 90 | 0·32 |
| 4 | 120 | 0·43 |
| 5 | 150 | 0·45 |

(*a*) On the grid below draw a line graph of average rate of sweat production against the duration of exercise.

Choose an appropriate scale to fill most of the graph paper.

(Additional graph paper, should it be required, will be found on *Page forty*.)

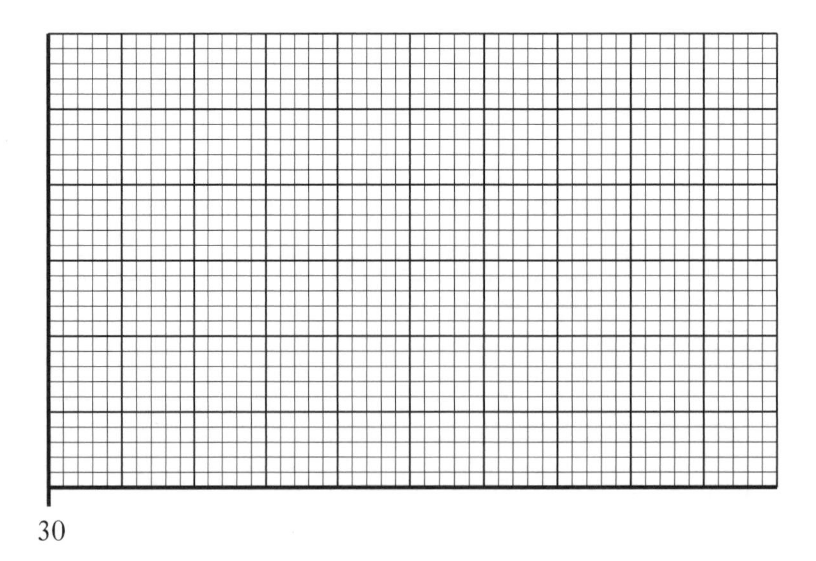


**2**

(*b*) (i) Give **two** variables, not already described, which should be kept constant to allow valid comparison of the exercise trials.

1 ______________________________

2 ______________________________ **2**

DO NOT WRITE IN THIS MARGIN

*Marks*

**12. (*b*) (continued)**

(ii) State how the procedure could be improved to increase the reliability of the results.

______________________________

______________________________ **1**

(*c*) Explain how the units of sweat production used in this investigation would allow a valid comparison between different individuals to be made.

______________________________

______________________________

______________________________ **1**

(*d*) Calculate the total mass of sweat produced per $cm^2$ during exercise trial 3.

*Space for calculation*

__________ mg per $cm^2$ **1**

(*e*) Predict the rate of sweat production which would be expected in an exercise trial with a duration of 180 seconds.

__________ mg per $cm^2$ per minute **1**

(*f*) (i) Sweat production is a corrective mechanism used in the regulation of body temperature.

Explain why maintaining body temperature within tolerable limits is important to the metabolism of humans.

______________________________

______________________________ **1**

(ii) Give the term used for animals which use changes in metabolism to regulate their body temperature.

______________________________ **1**

**[Turn over**

DO NOT WRITE IN THIS MARGIN

*Marks*

**13.** (*a*) The diagram below shows apparatus used in a water culture experiment to investigate the effect of the lack of magnesium on the growth of barley seedlings. The water culture solution provided all the elements needed for normal plant growth apart from magnesium.

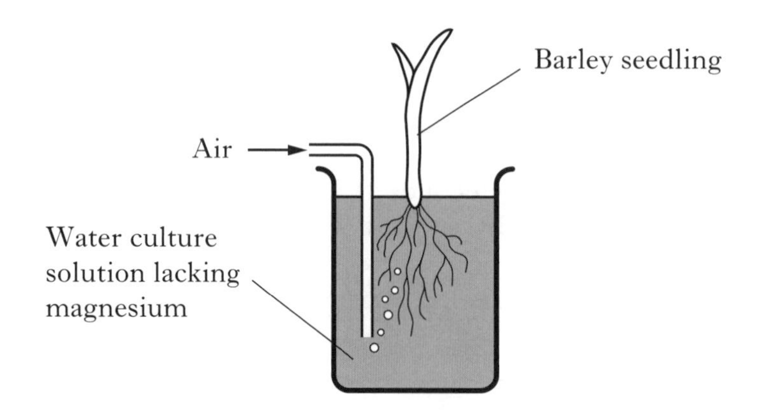


(i) Describe a suitable control for this investigation.

________________________________________

________________________________________ **1**

(ii) Explain why bubbling air into the water culture solution increases the uptake of elements by barley roots.

________________________________________

________________________________________

________________________________________ **2**

(iii) Other than overall reduction in growth, give **one** symptom of the deficiency of magnesium on the development of barley seedlings.

________________________________________ **1**

(*b*) State **one** role of iron in the growth and development of humans.

________________________________________ **1**

DO NOT WRITE IN THIS MARGIN

*Marks*

**14.** (*a*) The mating behaviour of red deer is influenced by photoperiod.

The graph below shows how the frequency of mating activity of the deer is related to photoperiod.

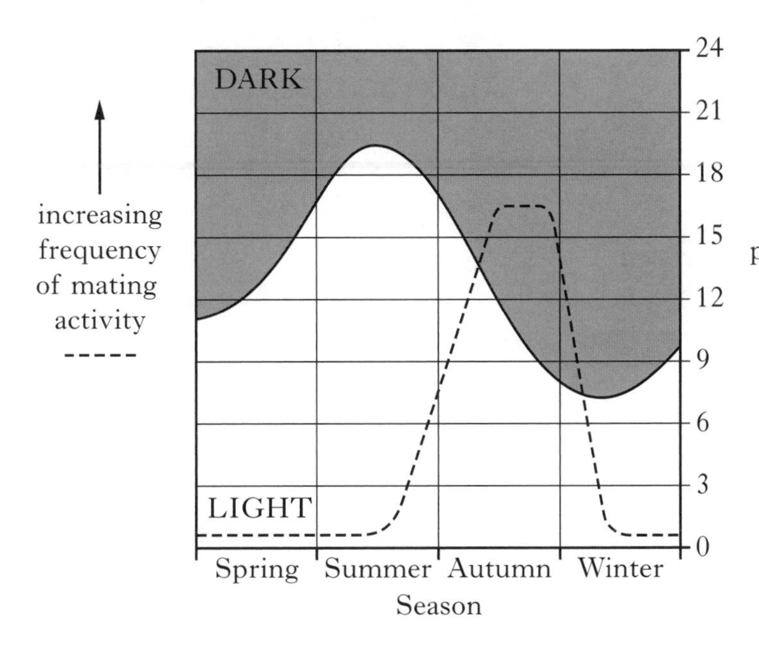


(i) Give the photoperiod at which frequency of mating activity first reaches its maximum.

________________ hours **1**

(ii) Suggest how the timing of mating behaviour is an advantage to red deer.

________________________________________

________________________________________ **1**

(*b*) (i) State the term used to describe the growth movements of plant shoots towards light.

________________________________________ **1**

(ii) State the environmental conditions under which a plant shoot would become etiolated.

________________________________________ **1**

**[Turn over**

DO NOT WRITE IN THIS MARGIN

Marks

**15.** The diagram below shows the changes in the plant communities present at various times after a field was abandoned.

| **Pioneer communities** colonise abandoned field within one year | **Intermediate communities** develop over 150 years | **Final community** established 200 years after the field was abandoned |
|---|---|---|

(*a*) Give the term used to describe the process of gradual change in plant communities shown in the diagram.

______________________________ **1**

(*b*) Give the term used to describe the final community in this process.

______________________________ **1**

(*c*) State **one** difference between a pioneer community and a final community.

______________________________ **1**

*Marks*

## SECTION C

**Both questions in this section should be attempted.**

Note that each question contains a choice.

**Questions 1 and 2 should be attempted on the blank pages which follow.**

**Supplementary sheets, if required, may be obtained from the Invigilator.**

**All answers must be written clearly and legibly in ink.**

**Labelled diagrams may be used where appropriate.**

**1.** Answer **either** A **or** B.

**A.** Give an account of mutation under the following headings:

(i) occurrence of mutant alleles and effects of mutagenic agents; **3**

(ii) gene mutations and their effects on protein structure. **7**

**(10)**

**OR**

**B.** Give an account of adaptations of plants under the following headings:

(i) adaptations of xerophytes for maintaining a water balance; **6**

(ii) adaptations of hydrophytes to their environment. **4**

**(10)**

**In question 2, ONE mark is available for coherence and ONE mark is available for relevance.**

**2.** Answer **either** A **or** B.

**A.** Give an account of the structure of chloroplasts and the carbon fixation stage of photosynthesis. **(10)**

**OR**

**B.** Give an account of the structure of the plasma membrane and the cell wall and describe their roles in relation to osmosis in plant cells. **(10)**

[*END OF QUESTION PAPER*]

**[Turn over**

DO NOT WRITE IN THIS MARGIN

**SPACE FOR ANSWERS**

DO NOT WRITE IN THIS MARGIN

**SPACE FOR ANSWERS**

DO NOT WRITE IN THIS MARGIN

## SPACE FOR ANSWERS

DO NOT WRITE IN THIS MARGIN

**SPACE FOR ANSWERS**

**SPACE FOR ANSWERS**

DO NOT WRITE IN THIS MARGIN

**SPACE FOR ANSWERS**

DO NOT WRITE IN THIS MARGIN

**SPACE FOR ANSWERS**

ADDITIONAL GRAPH PAPER FOR QUESTION 12 (*a*)

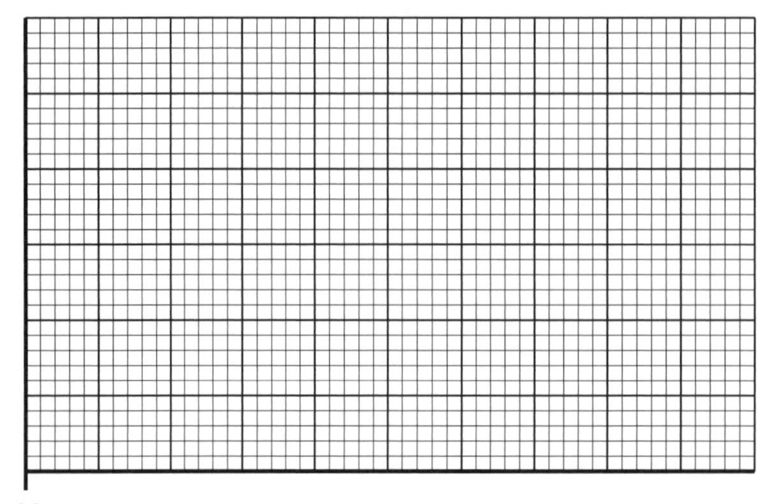